The Grant Writer's Handbook

How to Write a
Research Proposal and Succeed

The Grant Writer's Handbook

How to Write a Research Proposal and Succeed

Gerard M Crawley
University of South Carolina, USA

Eoin O'Sullivan
University of Cambridge, UK

Imperial College Press

Published by

Imperial College Press
57 Shelton Street
Covent Garden
London WC2H 9HE

Distributed by

World Scientific Publishing Co. Pte. Ltd.
5 Toh Tuck Link, Singapore 596224
USA office: 27 Warren Street, Suite 401-402, Hackensack, NJ 07601
UK office: 57 Shelton Street, Covent Garden, London WC2H 9HE

Library of Congress Cataloging-in-Publication Data
Names: Crawley, Gerard M., author. | O'Sullivan, Eoin, author.
Title: The grant writer's handbook : how to write a research proposal and succeed /
    Gerard M. Crawley & Eoin O'Sullivan.
Description: New Jersey : Imperial College Press, [2015] | Includes
    bibliographical references and index.
Identifiers: LCCN 2015032152| ISBN 9781783267590 (alk. paper) |
    ISBN 9781783264148 (pbk : alk. paper)
Subjects: LCSH: Science--Research grants--Handbooks, manuals, etc. | Research grants.
Classification: LCC Q180.55.G7 C73 2015 | DDC 658.15/224--dc23
LC record available at http://lccn.loc.gov/2015032152

British Library Cataloguing-in-Publication Data
A catalogue record for this book is available from the British Library.

In-house Editors: Catharina Weijman/ R. Raghavarshini

Typeset by Stallion Press
Email: enquiries@stallionpress.com

Printed in Singapore

Foreword

By William C. Harris,
Founding Director General of Science Foundation Ireland

Investment in research and advanced education is rising rapidly around the world, not only in established 'knowledge-based' economies, but in many newly developing countries as well. A critical part of building an internationally competitive research and innovation system is the establishment of research grant funding programs based on peer (i.e., independent expert) review. Researchers need to learn how to effectively communicate to reviewers both the quality of their research and their own ability to carry it out. In addition, as new investments in research are driven, at least in part, by efforts to stimulate economic productivity and growth, researchers will increasingly be challenged to explain the value of their research in terms of industrial or socio-economic impact.

Connecting research to society (national security, medical treatment advances, new companies and employment) was an integral component in Vannevar Bush's 1945 report: *Science — The Endless Frontier*. The Bush report continues to provide the most effective rationale for investment by government entities in economic and national security for any state or nation. And it is important to note that the Bush report provided the framework to start Science Foundation Ireland (SFI) including one of our initial investments — where we borrowed the word 'frontier' from Bush — when we set up the Research Frontiers Program to support important ideas outside of our primary focus on the strategic priorities of biotechnology (BIO) and information and communications technology (ICT).

Importantly, the authors of *How to write a Research Proposal and Succeed* played key roles in establishing SFI and helped to develop its modern peer review system — essential for a 21st century research funding organization. This book reflects their experience, first hand knowledge and good humor as well. Rarely do you find this kind of experience in a treatise that is so important to modern countries and states. In fact the Crawley–O'Sullivan book represents the evolution of the competitive funding process that has its roots in *Science — The Endless Frontier*, the US National Science Foundation, Science Foundation Ireland and the European Research Council. This book brings to life a very important subject that is likely not well understood outside of the senior, funded researchers worldwide. And, the book is interesting to read and should be a must-read for any young researcher or government policy maker (state or country) determined to understand what is essential for success and to avoid waste and missed opportunities.

How does a country or state attempt to move from low cost, low skill manufacturing to more knowledge-intensive activities, including those involving significant research and innovation capabilities? Clearly an investment in a world-class education system is one necessary condition. Another is the need to improve the innovation potential of the country by encouraging high quality, competitive research within the country, and by promoting interactions between the academic researchers and the local and international business community. A country with an excellent education system plus a productive research community is very attractive to international companies particularly if the cost of doing business in the country is competitive.

The world has changed greatly in the past 20 years and wise and thoughtful leaders can enable countries and states to succeed and grow from low level manufacturing to state-of-the-art work and high value employment. Ireland remains the ideal case study for understanding how to do this given its evolution, its size and success. That is, Ireland is small enough to understand what is essential and what works, and is complex enough for large countries to learn from as well. When we started SFI in 2001, we advised the government that this transformation would likely take about ten years — given the magnitude of the task and the required culture change as well. In 2011, Ireland became the home of an IBM

Research Center — 10 years after the commitment to SFI. While SFI and Ireland provide a 'case study' on the importance of strategic focus and how to generate economic value from strong research programs, specific chapters in this book should be studied as well (e.g., Chapter Six: Partnerships and Chapter Seven: Impact).

Some of the countries that have succeeded in making the transition to a more knowledge-based economy have improved the quality of the research carried out within the country. But this required a change in how the universities functioned — at least the engineering and sciences in the institutions. Real performance metrics and new approaches to real partnerships were essential. Investment in research must be managed carefully by professionals who understand research and advanced education and are trusted by industry. To establish successful industry-university partnerships — to advantage economic prosperity — an independent entity like SFI is essential: that is, an intermediary organization is essential to measure and hold accountable the entities for results and not hype.

The authors succeed in providing an excellent 'how to' guide for researchers preparing proposals for competitive peer review. They draw on their own experience of reviewing and organizing reviews in a number of different countries, both those with a tradition of competitive peer review as well as some that are just beginning the process. In addition, they have made use of their colleagues in many countries to provide their insights and advice. The book is filled with excellent examples, both positive and negative, plus numerous quotes from reviewers that illustrate many of the points that the authors make.

While the book would be especially valuable to researchers at an early stage of their careers, more experienced scientists and engineers will find the book beneficial, especially if they have not had much experience with peer review using international reviewers. And although the emphasis in the book is on proposals in the sciences and engineering, proposal writers in many other fields will find the book useful.

One other point that should be made about this book is that while it is intended as a guide to writing a successful grant proposal, it also contains much valuable advice on 'best practices' in carrying out research successfully. Sections on managing people, managing budgets, fostering clear and regular communication, keeping up with advances in the field, being open to interdisciplinary activities, and preparing for future proposals will all help young scientists (and maybe some older ones) to continue doing outstanding science.

This book should also be useful to the staff of the research agencies in a country trying to establish a competitive research funding organization. Many of the staff of these agencies who have had little experience with organizing peer reviewed competitions would benefit from some of the insights provided in this volume and, importantly, engaging individuals with substantive experience as colleagues in the early years to avoid missteps.

The book should also be helpful to the administrators in universities and research institutions who are charged with encouraging their staff to submit research proposals to their respective funding agencies, especially if these agencies are now employing a competitive peer review process using international reviewers. Finally, this book should be required reading for policy makers in any state or country that is determined to move its university system into becoming far more competitive in the 21st century and to work in new ways to advance the country. This book is much needed and timely.

Acknowledgments

The authors would like to thank the following colleagues who agreed to be interviewed for 30 minutes or more about their experiences as reviewers and/or grant-writers, who read and commented on drafts of various chapters and who gave permission to use sections of their proposals as examples. Some also provided paraphrased and redacted versions of reviews they had written which provided us with specific examples.

These are all busy people and we very much appreciate their professional and personal input in preparing this volume. The list, in alphabetical order, includes:

Kamal Abdali (Computer Science), Swann Adams (Public Health), Claudia Benitez-Nelson (Marine Science), Frank Berger (Biology and Medical Sciences) Anthony Boccanfuso (Chemistry), James Burch (Public Health and Epidemiology), Ronan Daly (Nanotechnology), Michael Felder (Genetics), Joseph Finck (Physics), Stephaine Frimer (Education), Frank Gannon (Biology), Giorgia Giardina (Engineering), Charles Glashausser (Physics), Bill Harris (Chemistry), Thomas Healy (Chemistry), James Hebert (Public Health and Epidemiology), Richard Hirsh (Computer Science), Rory Jordan (Physics), Ohad Kammar (Computer Science), James Kellogg (Geological Sciences), Mary Kelly (Biology), Peter Kennedy (Engineering), Scott Little (Chemistry), Graham Love (Biology), Aisling McEvoy (Materials Science), Timothy Mousseau (Environmental Science), Yukihiro Ozaki (Chemistry), Daniel Reger (Chemistry), Roger Sawyer (Biology), Shirley Scott (Education and Humanities), Susan Steck (Public Health), Marko Tainio (Epidemiology and Environmental Health),

Michael Thoennessen (Physics), Vicki Vance (Genetics), Alan Waldman (Biology), Ken Wilson (Economics) and Jennifer Yip (Public Health).

Finally, the authors would like to give special thanks to their families for their support and advice.

Glossary

The usage of certain terms is different around the world even in countries that speak the same language like the United States, Canada, the British Isles, Ireland and Australia. Therefore we thought it useful to list a number of such terms that are used throughout the book and explain the meaning to an audience that may be unfamiliar with some of the particular usage and abbreviations used in the book.

Call for proposals (CFP): Or sometimes simply 'the Call' or the 'Call document'. This may also be called the **request for proposals (RFP)**. This term refers to the material released by a funding agency to invite the submission of proposals. The information is normally posted on the website of the funding agency and explains the requirements for submission including a deadline, if any, page limits on parts of the proposal or on the overall proposal, the amount of the awards, the criteria that will be used to evaluate the proposals plus any other information required to be submitted as part of the proposal.

Co-investigator (Co-I): Or sometimes a Co-PI. A Co-I is a researcher who leads a particular strand or (sub-) project within a grant. Typically a Co-I will have helped the 'principal investigator' (see below) develop the grant proposal. The Co-I's research track record will be an important feature of the proposal.

Cost sharing: Cost sharing refers to an arrangement where an organization other than the funding agency agrees to fund part of the research project. For example, an industry partner may pay for 25% of a project relevant to their research agenda. That 25% would be considered the 'cost share' and may be in cash or 'in-kind' contributions, (e.g., the equivalent cost of the time of one of their employees working on the project, or cost of access to their equipment used in the project). Many funding agencies have industry-relevant programs where some level of cost share is required. When an agency requires partners to the costs of a project equally these are referred to as **matching funds**.

Curriculum vitae (CV): Also sometimes called a **resumé**. This describes the academic history either of a person submitting a proposal or a collaborator on a proposal. The information normally required includes the degrees awarded, including the date and institution, relevant positions held including the person's current post, the academic papers and books published, a list of talks presented at conferences, patents awarded, and previous grants obtained. In many cases, there are limits on the length of the CV and/or there may only be a specific number of the most important and relevant papers required.

Direct costs: Direct costs are readily identifiable costs incurred by a specific project. The main categories of direct costs include (but are not always limited to): the salaries (and employment benefits) of research team members, equipment, materials/supplies/consumables, travel (and accommodation). See also **indirect costs**.

Funding agency: A funding agency may be a government body funding a range of research domains (e.g., a National Science Foundation or Research Council), a government mission agency funding strategic research priorities (e.g., agencies of a Department of Health, Defense, or Energy), or private organization (e.g., a charitable foundation). Funding agencies typically provide research grant funds for a defined period to a researcher's host institution for that researcher carry out projects described in their grant proposal.

Grant officer/program officer: The grant officer is a funding agency's designated official responsible for managing aspects of a particular grant. The grant officer is responsible for the review, negotiation, award, and administration of a grant. Officials who have overall responsibility for particular funding initiatives or research grants in a particular research domain are called Program Directors or Program Managers. Program Directors may also be grant officers for particular awards.

Hypothesis: A hypothesis is speculative statement relating to an observed phenomenon which should be formulated in such a way as to be verified or falsified by research analysis or experimentation.

Indirect costs: (or **Overheads**, see further) (Sometimes, in the UK this is called **full economic costing**). In contrast with **direct costs** (see previous page), indirect costs are those institutional expenses incurred by hosting research institutions and which are difficult to directly attribute to a particular project. Examples of indirect costs include: heating and electricity supply to buildings, building security, cleaning and maintenance, institutional administrative services, such as accounting, IT support, payroll, human resources, library expenses, etc.

Literature review: A literature review (or 'lit review') is an overview of previously published research relevant to the proposed research. In particular, the lit review should describe the state-of-the-art in the relevant research domain(s), highlighting how the proposed research relates to what is currently known, is distinct, novel, methodologically sound and has the potential to advance knowledge in the field.

No cost extension: If there are funds remaining at the end of the grant period, most funding agencies will extend the time over which these funds can be expended. This is referred to as a 'no cost extension' since no additional funds are made available. In some cases, approval for extending the time for spending grant funds must be requested explicitly. In most cases, however, there is a policy of a standard period of time, six months or even one year, during which unexpended grant funds can be used for the purposes designated in the original award. In such cases, this policy will be stated in the web site of the funding agency.

Overheads: An overhead is a charge levied on an award and paid to the institution supporting the award to help cover the 'indirect cost' or "full economic cost" (see above) of the research incurred by the institution. This may include the cost of providing space, with its attendant heating, cooling and cleaning costs, providing benefits to personnel hired to carry out the research, library costs, and financial accounting. In some cases, the overhead rate is negotiated with a given institution by one agency but will be used by all government agencies in dealing with that institution. Overhead rates may be negotiated on a multi-year basis. In most cases, overhead is not charged on equipment purchased by an award.

Preliminary proposal (or 'pre-proposal'): In cases where many proposals for a competition are anticipated or where the amount of the award is large and an extensive review process is required, there may be a preliminary proposal or pre-proposal required. Such pre-proposals are normally much briefer than the full proposal and are therefore easier to review. Only the authors of pre-proposals which pass the first review are asked to submit a full proposal. The aim is to simplify the proposal review process and to avoid having reviewers read large numbers, or large numbers of pages, of full proposals.

Principal investigator (PI): Strictly the term principal investigator should probably be restricted to the person (or persons) responsible for the administration and leadership of an award. In many cases, however, including in this book, the term principal investigator is used also to mean the **applicant for a grant** or the **author of a proposal**. Most funding agencies prefer to deal with a single person in administering the award. In cases where there is more than one key researcher in a proposal these individuals are often designated as 'Co-Investigators' (see above). For some types of non-research grant (e.g., workshop grants, education and outreach grants), some funding agencies use the term '**project director**' or '**project manager**' rather than PI.

Quote for equipment: This refers to the price asked by a seller for a particular piece of equipment or for a service. Such a quote is almost always presented in writing. Such a 'quoted' price, often with an expiration date, can be used to obtain a firm estimate of the cost of particular equipment or services when presenting a budget.

Reviewer: A person who reviews or assesses a proposal. In most cases more than one reviewer will be asked to assess any one proposal and there is normally a trade-off between the number of reviewers used and the cost of the review process. Reviewers are normally asked to provide their assessment in writing using the criteria listed in the CFP, as well as providing an overall score or alpha-numeric rating of the proposal. Reviewers are expected to have some expertise in the subject matter of the proposal. Reviews from expert reviewers are often a valuable input to the applicant of a proposal even if the proposal is not funded and in many cases are shared anonymously with the applicant.

Work package: A work package is a term that is sometimes used to describe a 'sub-project', task, or set of tasks within the overall project. Work packages are used to help manage more complex projects by breaking them down into different 'mini projects' or activities with their own deadlines, deliverables, staff responsibilities, etc. This helps the PI oversee progress and manage the grant more effectively.

Contents

Chapter One

Introduction

You have to learn the rules of the game. And then you have to play better than anyone else.

— Albert Einstein

This book is designed as a guide to writing grant proposals, aimed at researchers with varying levels of research grant experience. Although primarily aimed at early career researchers applying for their first grant, this book will also be useful for postdoctoral researchers helping their supervisors prepare proposals, as well as researchers in countries with emerging research funding systems that do not have a long history of competitive peer review.

In particular, the book shares important insights, experiences and practices which can help grant applicants communicate the novelty, strengths and importance of their research ideas. It also provides researchers with tips for how to convey their competence to plan, manage and deliver on their research program.

The book takes readers through the entire grant application cycle, from generating the initial research ideas to formulating the research question, drafting (and redrafting!) the proposal, the review process, and responding to reviewers' comments. All the main sections of a grant application are considered in detail, including how to write an effective abstract, provide the right level of technical detail in the methodology section, effectively highlight the strengths of the research team, design a convincing budget, and explain the future impact of the research.

Important lessons are illustrated with real examples from successful (and unsuccessful!) grant applications. Key messages are emphasized with quotes from reviewers, and there are insights from funding agency officials from a range of organizations around the world. This guide to 'grantsmanship' aims to share insights into the thinking of experienced reviewers: what appeals to them and what turns them off, what is likely to encourage them to either recommend funding or decline a proposal.

1.1 Why This Book is Needed?

The skills and knowledge needed to write a successful grant proposal are not the same as those required to be a good researcher. Being a good researcher is a necessary but not sufficient condition for writing a convincing grant application. Typically, the additional skills and insights required to write a competitive proposal are learned through experience, rather than being taught by research mentors or shared by research administrators.

To address these challenges, this book collects and shares important tips, useful examples and effective practices to help in writing a grant proposal that can persuade individual reviewers, review panels and funders that your research idea, track record and project plan are worth investing in.

Developing a compelling grant application requires an ability to engage the reviewers. It should convey the excitement of the research idea and its potential impact, while setting it in the context of the broader research field. A strong application will grab the readers' attention from the beginning of the proposal, and maintain their interest and excitement throughout the document. A successful grant proposal will also persuade the reviewers that the applicant has the ability to *manage* his or her grant. It should make clear that the applicant has the capability and competence to build an effective research plan, manage the research team and resources, and deliver on the project's goals. So while this book is primarily intended as a *'how to'* guide for writing grant proposals, it does offer some effective practices for managing research projects as well.

The book captures some of the most important grant-writing techniques and practices from different countries that have established, sophisticated research systems. Over the years, researchers in these countries have learned many lessons about how to effectively communicate their research ideas to reviewers. Experienced and successful grant-writers will have developed an intuition about the right level of technical detail to include, what annoys reviewers and what convinces them.

The book also gathers insights and advice from countries with emerging research systems, especially those that have only recently adopted competitive peer review-based approaches for allocating research funding. Researchers in these countries may have to develop new approaches for communicating the quality, novelty and impact of their proposed research. In particular, this book shares key messages from international experts who have been involved in assessing grant applications or managing new research grant competitions in such countries, and who have therefore seen feedback from reviewers at first hand. These experts are well placed to identify some of the most common mistakes and misconceptions made by researchers applying to peer-reviewed grant competitions for the first time.

Investment in basic and applied research is rising rapidly around the world, not only in established 'knowledge-based' economies, but in many middle-income and developing countries. Many of these investments are driven by efforts to stimulate economic productivity and growth. Consequently, researchers may increasingly be challenged to explain the value of their research in terms of industrial or socio-economic impact. This book also provides guidance on how to convincingly communicate the planned outputs, outcomes and future impact of a research project.

1.2 Who is This Book For?

This book is aimed primarily at early career researchers applying for their first grant. It will also be useful for other researchers, from postdoctoral researchers helping their supervisors prepare proposals, to more established researchers in countries with less experience of competitive peer review-based funding programs.

The book focuses on sharing practices for applying to funding agencies supporting research in scientific and technical disciplines, and learning lessons from other proposals. In particular, the book includes examples from the *natural sciences* and *engineering*, as well as medicine, the social sciences, economics and management. We have not included examples from the humanities and the arts, although many of the key messages will be pertinent to these fields.

The experience of the authors is that many of the proposals we have reviewed, especially from younger researchers or researchers in countries with emerging research systems, lack some of the fundamental concepts of what constitutes a good proposal and include many basic mistakes in a proposal prepared for peer review. We believe that this book will provide a service to grant proposal writers in trying to help avoid many of the potential pitfalls in preparing a proposal.

In addition, the authors believe that the book will be useful to the staff of the research agencies (especially recently established funding organizations). Many of the staff of these agencies may have little experience with organizing peer-reviewed competitions as a means of allocating research funding, and we believe that they will benefit from the insights provided in this volume.

The book should also be helpful to the administrators in universities and research institutions who support staff submitting research proposals to funding agencies, especially if these agencies are now using competitive peer-review processes involving international reviewers. Staff at new universities or other recently established research institutes — i.e., those which have not had much time to develop a 'corporate memory' of what makes a strong proposal or what are the most common grant-writing mistakes — should also find this book useful.

1.3 What's in This Book?

This book shares experiences, insights and practices to help grant applicants effectively communicate the novelty and strengths of their research ideas in funding proposals. Key messages are emphasized with quotes from the experts we interviewed in the course of writing this book, and we paraphrase and synthesize quotes from reviews of actual proposals. We also illustrate important practices, with examples or 'stories' (both positive and negative) drawn from real grant applications. Although these examples have been modified to protect the identities of the original principal investigators (PIs), the lessons learned and the effective approaches they demonstrate have not. We hope that these examples, drawn from real life experience, will make the discussion more alive and meaningful for the reader.

Some of the quotes from experienced reviewers, the lists of 'dos and do nots' and the examples of good practice highlighted in this book may seem like common sense. However, the authors have seen enough proposals and reviews to know that the same mistakes are made over and over again. Even the very best researchers can make the most elementary grant-writing errors, and are perfectly capable of boring, annoying or confusing their reviewers.

In collecting these lessons, practices and 'stories', the authors have drawn upon their considerable experience in setting up grant competitions, managing peer-review processes, and reviewing proposals from a wide variety of countries (with research systems at different levels of maturity). We have also made use of our many contacts with successful grant writers in different research domains of the physical and biological sciences, engineering, social science, economics and management to obtain their insights about what makes a good research proposal in their particular field (and what issues they have seen that cause proposals to be unsuccessful). These interviews involved in-depth discussions with over 30 experienced reviewers and funding agency officials in several different countries. We also received significant input from younger researchers across a range of disciplines to better understand which aspects of the grant application process they struggle with and where they feel guidance would be most useful. In practice, we found general agreement on what makes a good proposal, independent of the home country of the reviewers and even, to some extent, of the discipline involved.

The book systematically goes through each step of proposal preparation and submission, from generating the initial research ideas, to formulating the research question, drafting *and redrafting* the proposal, understanding the review process, and responding to reviewers' comments. Effective practices on how to approach all the main sections of a grant application form are considered in detail, including: how to write a compelling abstract; construct a literature review of the 'state-of-the-art'; provide the right level of technical detail in the methodology section; effectively highlight the strengths of the research team (and collaborators); design a convincing budget; and explain the future impact of the research.

The careful reader may note that there is a certain amount of repetition in the book — the same advice is given more than once. Partly this is because we anticipate that the chapters in the book may well be read independently for help in specific areas. But even more importantly, our experience with teaching has taught us that repetition is often a very valuable tool in conveying a message.

1.4 How This Book is Organized?

The book is structured into 12 chapters plus appendices. In the next chapter, **Chapter Two**, we deal with the 'main idea' behind a proposal. As well as stressing the importance of having an excellent concept and research question to address, we offer some advice on how to generate ideas and keep current with your field.

In **Chapter Three**, we discuss the actual review process itself, since we believe that an understanding of the process will help the writers of proposals do a better job in actually presenting the material.

In **Chapters Four to Nine** we deal with various important sections of the proposal, including, referencing, the budget and the body of the proposal.

In **Chapter Five**, the authors of proposals are encouraged to allocate sufficient time to allow redrafting the proposal using friends and colleagues as critical readers. There is also a discussion of more complex proposals involving partnerships in **Chapter Six**.

Chapter Ten gives advice on how to respond to reviewers' comments convincingly.

Chapter Eleven discusses approaches to applying for special kinds of proposals, e.g., for travel or equipment, or for early career investigators.

In **Chapter Twelve** some advice is presented on managing an award (and planning the next one), assuming that the proposal is funded. The ultimate goal is, of course, to continue to obtain research funding throughout a long and successful career.

Appendix One presents an outline of how to run a competitive grant competition. We anticipate that this should be helpful not only to prospective authors of proposals, but also to funding agency staff and the staff of research offices in universities and institutes.

Finally, **Appendix Two** provides a list of websites where international funding agencies provide advice on proposal preparation.

Chapter Two

The Research Idea

The true sign of intelligence is not knowledge but imagination
— Albert Einstein

Εὕρηκα! [*I have found it!*]
— Archimedes

So here you are with fingers poised over the keyboard of your computer. It's time to write that proposal to fund your new project. But where to start?

Any good proposal starts with an idea. If the author of a proposal does not have a good idea, no matter how polished the finished product, the proposal is not likely to have great success either with reviewers or with the agency providing the funding. This is the area where the author's experience, knowledge of the subject matter, creativity and imagination most come into play.

2.1 What Makes a Good Idea for a Research Proposal?

Words that are often used in a call for proposals (CFP) from a funding agency, to describe the concepts they are looking to fund, are *original*, *significant* or even *beyond state-of-the-art*. Funders are looking for fresh ideas and not merely the 'same old' approaches that have been tried before.

> *What I look for in a good proposal is excitement.*
> *A proposal that is simply incremental turns me off.*
>
> **— Quote from an experienced reviewer**

Funding agencies want the projects they fund to make a difference to the development of the field, possibly leading to new directions or to solving a problem. Such problems exist in all fields, and finding ways to solve them is at the core of a good research proposal.

Reviewers are often uneasy about an applicant who simply wishes to continue his or her PhD project in a new environment. Distancing oneself from a thesis advisor and being seen as an independent researcher with new ideas can be very important.

An example of moving in a new direction and not settling for incremental advances.

One of the scientists we interviewed described a situation where his group had shown how an estrogen receptor switched a particular target gene to 'on'. This was a major breakthrough and led to a number of papers in high impact journals. However, once the first two or three papers were published, the group decided that they would move on to another area and not to continue simply finding more examples of the same phenomenon. They believed that this would just be incremental work. Instead they moved into a new area of DNA methylation, and continue to make contributions in this area.

Firstly, you will need to read the CFP carefully to make sure you are addressing the requirements of the funders. There is no point submitting a proposal with an exciting idea if the funding agency is looking for the solution to a particular practical problem. For instance, funding agencies (particularly those in countries with emerging economies), often require that the proposals they fund make a positive difference to the economy of their country. Therefore, it is critical in your proposal to address the concerns of the agency, and to look for ways in which your project is likely to have an impact beyond gains in knowledge. We will discuss further the importance of the impact of research in Chapter Seven.

I have a number of problems with this proposal. First, where is the novelty? IR spectroscopy is a well-known analysis tool with many practical applications. Perhaps the best known is the measurement of blood alcohol content. Commercial instruments are readily available. What will be new in the current project?

— Comment from a reviewer

2.2 Where Do Ideas for a Research Project Come From?

Students finishing a thesis and moving to a research post are often tempted to repeat their thesis project under a slightly different guise. This may be a tempting and even reasonable strategy, where there remain interesting aspects of the subject to be explored. However, in most cases it is wise to strike out in a new direction when looking for a project. The best projects are those that capture your interest. Chances are that other people will also find this to be true, and your enthusiasm can be an added benefit.

However, no matter how captivated you are with a topic, it is always a good idea to ask yourself some important questions, such as:

- What difference will the results of the project make to the overall understanding of the field or of a neighboring field?
- How significant will the result be, even if the project is successful?
- Who will be interested in the results? People in my field? People in neighboring fields?
- Can the project be carried out with the equipment and techniques available now?
- What positive impact will the project have on society?

People will be interested in your results if they increase understanding in the field or if the project has the potential to help solve a problem locally or globally.

Sometimes an idea may be extremely important, but the technology is not yet available to be able to carry it out. **Timing can be critical**. Choosing a project that can be carried out successfully now, but which would have been impossible previously, is usually a good strategy. In some cases, the development of the technology itself may be a good research topic. Often such technical developments can spin off into numerous projects. Keeping abreast of current technology is also important in allowing one to capitalize on new developments.

Another reason for being familiar with the technology in your field is that it may turn out to apply to a problem in another field. There are many examples where a technique developed in one science has made an enormous impact in another. One example is the use of mass spectrometry to measure $^{13}C/^{12}C$ ratios, which has allowed geologists to study ancient climate patterns.

Archimedes gets an idea for a grant proposal.

So, as your career advances, it is important to continue to keep abreast of the latest advances in the field. The most common mechanism is to **read the relevant journals regularly**. With most journals available online, it is easier than ever to keep in touch with current work. In most fields of research, there are half a dozen or more journals that contain reports of the most important research. It is important to make time to read these critical journals to make sure that you know what is currently happening in your field of interest. There are also more general 'popular' magazines containing summaries of interesting areas of current research. In physics, for example, the popular journal *Physics Today,* produced monthly by the American Physical Society (APS), summarizes interesting new results in different areas of physics. Professional groups in other countries produce similar overviews for a more general audience.

As you read, collect and store ideas for possible proposals.

One researcher we interviewed, a chemist, told us that as he reads the literature, he collects ideas on index cards. Even when he took his first academic position, he had already collected a small box of cards with ideas. He still does this today, even though he is now an experienced researcher with a strong record of funding. Of course, there are modern electronic methods of keeping track of good ideas, but the concept is the same.

Another advantage of reading the literature regularly and carefully is that it can prevent your spending needless time on projects already published. Most scientists have experienced being 'scooped' on a project where a result is published just as they were preparing their own for publication. This is unfortunate but in most cases unavoidable. What is avoidable is spending time on a project already reported in the open literature. With modern electronic search tools, this should now be very rare indeed.

An example of the importance of reading the literature from one of our interviews — much time would have been saved by checking the literature.

When Dr. B was a postdoc at a large National Laboratory, the group with which he was working was developing an array of Barium Fluoride detectors. They discovered that if they cooled the detectors to zero degrees C, the response of the detectors was much better. They wrote up this observation for submitting to a journal. However, as they looked at the literature, they discovered that this observation had already been published. If they had been more conscientious about reading the literature, they might have saved themselves a good deal of time.

Ideas for research projects often come at unexpected times even when you are relaxing, maybe in an *onsen* (hot spring) or on a long airplane flight. One senior Japanese scientist has said that many of his best ideas come when he is daydreaming during a boring committee meeting. On the other hand, it is useful to deliberately create situations to encourage stimulating ideas.

Often a group of researchers will form a **journal club** where each member brings a different journal article to the meetings, and explains it to the rest of the members. The subsequent discussion may not only lead to a better understanding of the current topic, but also generate interesting new research possibilities to explore. A journal club with student members has the added advantage of providing the students with the opportunity to present ideas to an informed audience. This can help improve their communication skills.

The notion of a journal club also emphasizes the importance of **communicating with colleagues**. With very few exceptions, most people find that their thinking is stimulated by discussions with other people. This can happen informally, for instance, by dropping into a neighboring office. In some departments or institutes, there are also formal arrangements to encourage interaction. For example, we have experienced some academic departments host afternoon coffees for the whole department with all of the faculty and students strongly encouraged to attend. The blackboards around the room can be filled with symbols as small groups discuss various topics. This is a tremendously stimulating environment, and we believe it serves as an excellent model for a vibrant research group in any field.

Many departments and research institutes also hold regular **seminars and colloquia**, where visitors from outside the organization are invited to present their latest research. Both the presentation and the subsequent question and answer session are often stimulating sources of new ideas. Unfortunately, budget problems or simply geography can make it difficult to invite outsiders. If institutions are close enough, in a large city for example, it is sometimes possible to share the cost of a speaker and the staff of the different institutions can all attend the presentation.

In addition, it is important to **attend conferences** related to your research field. National and international conferences are held regularly in many countries and you should take every opportunity to attend at least one such conference a year if possible. Not only will you have the opportunity to hear experts address current topics, but you will have the opportunity to meet and mingle with other participants. This can be both stimulating and informative. In the United States, for instance, there are numerous Gordon Research Conferences held every year in the various fields of chemistry. These conferences are designed to be very informal.

The number of participants at any given conference is limited, no minutes are kept nor are the proceedings published. The afternoons are kept free to permit informal discussion among participants. Similar informal conferences, are held in many countries of the European Union, and could serve as a model for countries that are trying to improve their research performance. In some cases, funding agencies have special travel awards to permit attendance at such conferences (see Chapter Eleven).

Interacting with colleagues in neighboring fields can also be very stimulating. In some cases, there may be new techniques that may be applicable to your field. Or concepts which are well developed in another field may be transferable to yours (and vice versa). Group theory was well studied by mathematicians before being applied to particle physics, while the use of statistics and large databases is in the process of revolutionizing biology.

Enlightened chairs and department heads will often make funds available to allow their staff to attend conferences to keep them up to date with current research. This can be especially valuable for young researchers who do not yet have their own research funding, and for staff who may be suffering a temporary hiatus in funding. The sensible leader must simply be cautious that this opportunity is not abused and some outcome should always be expected.

Finally, in exploring opportunities for new interesting ideas, the **stimulation provided by students** can be very important. Even students in elementary classes sometimes raise questions that can force one to examine a topic in a different way, and this can lead to an interesting research project. In graduate classes or seminars this possibility becomes even greater.

DO read the current literature in your field and keep track of research ideas that arise.

DO consider forming a journal club with colleagues and students.

DO discuss your ideas with colleagues.

DO attend seminars and colloquia in your own and neighboring institutions if possible.

DO attend national and international meetings as funding and time allow.

2.3 What Makes a Good Research Topic?

Selecting an appropriate research topic is one of the most challenging and important issues in writing a research proposal. But this is also one of the most difficult areas in which to give advice. Different fields of research have different cultures and approaches, and you need to be familiar with the particular research culture in your own domain. Certainly all your professional preparation is important, as is using all the tools discussed in Section 2.1 to keep you up-to-date with important issues.

There are numerous factors to consider in deciding on a research idea for your proposal. It should be **original**, **interesting**, **exciting** and **not incremental**. Ideally, the reviewer on reading your idea should ask himself or herself, 'why didn't I think of that topic?' The subject must be consistent with the goals of the funding agency, as spelled out in the CFP. And, if the CFP requires it, links to possible applications and with industry need to be established.

Some general approaches that apply to many proposals are:

- Providing new knowledge that moves the field forward.
- Providing a method of solving an outstanding problem in the field.
- Developing an improvement in the current technology and indicating the resulting new opportunities.

Providing new knowledge

A survey of the literature, discussed in Section 2.3, will place your topic in context and show how it might advance your field. There are always important questions in any field that beg for answers. However, you need to be familiar with the current status of your field to be able to ask these questions, and to decide on a strategy to answer them.

> *If the project is not going to move the field forward, reviewers will not likely recommend funding. If it does not have a broader significance, it simply becomes a hobby.*

In some cases there may be a controversial issue requiring clarification. If you can think of a research question that you can ask with an answer that might resolve this issue, you have a good basis for a proposal. Even better is the case where the question you pose can challenge some fundamental concept in the field, but this is a more onerous task. The burden of proof will be much more difficult, and you must be certain that your approach has not been tried before and that you are not wasting your time.

The goals should be ambitious, but not too ambitious

There is a natural tendency, especially for researchers without much experience in writing proposals, to want to solve the major problems in the field. They will choose broad topics that outweigh the resources, both in personnel and funding, that are available. Normally the CFP for any particular competition for funding will give an indication of the magnitude of the budget available, and this will give the proposer a good idea of the possible scope of a project for this competition. There is a major difference in expectation between a large 'center-type' grant involving many people, and a single investigator award.

This is not to say that the author of the proposal should not be ambitious in his or her proposal. Reviewers normally like proposals that reach into new territory and attempt difficult and challenging projects. If, however, a project is too ambitious and tries to do too much with available resources and time, reviewers will usually judge that the author is inexperienced or has poor judgment and they will mark the proposal down. A limited project that has an excellent chance of success can provide the basis for future proposals building on the success of the current work.

> *A common input from a number of reviewers was that many of the proposals they reviewed, especially from young investigators, were far too ambitious. The investigators tried to do too much with the resources of time and money that were available.*

It is also important to ensure that the research question you pose can be answered with the equipment available or that you can develop the technology to answer the question. This development may be a major part of the proposal, and you will need to face the question of what to do if the technical development stalls or is not possible. Not every project will succeed, but you need to convince the reviewer that your approach has at least a reasonable chance of success.

An example of the danger of choosing an overly ambitious research idea, *viz.* nanoscience.

One of the current exciting areas in physical sciences is nanoscience i.e., science at the scale of nanometers (10^{-9} m). This is approximately the molecular scale and has relevance to a wide variety of topics, from delivering medicine to specific organs, to manufacturing. As a result, it has become fashionable to wish to participate in this rapidly advancing area. However, many large laboratories in many countries are engaged in research in this area, so that attempting to start a research program in nanoscience, unless there are very large resources available, is very likely to be doomed to failure. If resources are limited, it is usually more advantageous to choose an area that is less competitive and in which fewer people are working.

What does it mean for the research idea to be 'original'?

All true research must be original and this is a common requirement in many, if not all, CFPs. Certainly you need to be sure that you are indeed doing research and not simply repeating work done previously. Nothing will kill a proposal more quickly than the reviewer realizing that the work you are proposing has already been done, or is simply derivative in nature. As mentioned earlier, choosing a topic that is both **original** and **important** and **NOT incremental** is definitely an advantage.

The importance of originality in formulating your research question — coming up with an original approach can make a big difference to the reception of your proposal by reviewers.

Dr. F, who we interviewed, described a proposal that he reviewed some time ago. The proposal was from a young person at the beginning of her work (she has subsequently had a great career). She proposed to study the effect of alcohol use by means of drosophila (fruit flies). She proposed to select drosophila that were either more or less susceptible to alcohol. To do so, the principal investigator (PI) took a heterogeneous group of flies and placed them in a flask and tubing that she had designed containing alcohol. When the fruit flies became intoxicated they fell to the bottom of the tube and were readily collected. By this means, she was able to screen out mutants that were more or less resistant to alcohol. Some of these same genes turned out to be present in humans. It was a very clever but simple idea and worked really well. The reviewer gave it a good score, and he turned out to be right. It helped the PI get her career get off to a great start.

There are, however, cases where the unique situation in a particular country might justify using well-known techniques that have worked in other places, now applying them to a different environment. The availability of family data over many, many generations in Iceland or the Republic of Kazakhstan for example, permits interesting genetic studies, even though the techniques were developed and used elsewhere. Similarly, the nuclear disasters at Chernobyl and Fukushima have given rise to interesting environmental studies using well-known principles, now applied to very different and unique situations.

An example emphasizing the 'local' originality of the project.

Dr. S, who we interviewed, had read a proposal from researchers in a developing country with a higher than normal incidence of women's deaths from breast cancer, and with a low rate of women seeking screening for breast cancer. They recognized the link between these two facts and sought to address them. They knew of a project in another country that had established and successfully addressed the complex cultural factors that prevented a segment of the female population from seeking breast cancer screening even when it was readily available. The researchers in the developing country saw parallels between the situation of the women of their country and the situation of women in the other country, and proposed to adapt the intervention to their own cultural situation. They set up a careful, comprehensive, and methodical project to address the factors that disinclined women to get screening for breast cancer. They were able to encourage women from a young age to undergo screening. The 'originality' of this project was simply in recognizing how to replicate and adapt a project that had succeeded elsewhere in order to meet a local need that had never been addressed. It was a very well developed proposal, and therefore was ranked highly and ultimately funded.

Therefore, it can be important to examine any unique characteristics of your local environment in looking for appropriate research questions. You may be able to make use of techniques available elsewhere, but apply them to a unique local situation to craft an original project which has local value and possibly even wider applicability.

Hypothesis-driven research

A hypothesis is formally defined as a tentative explanation accounting for a set of facts that can be tested by further experimentation. In other words, hypothesis-driven research asks a question that can be answered unambiguously by experiment. Features of a good hypothesis are that it should be **simple**, **clear** and **testable**. It should also lend itself to constructing a set of experiments that will prove it either true or false. Good hypotheses should **not** include words like 'may' 'might' or 'could' since these make the statement impossible to falsify (i.e., show that it is untrue). Nor should you include words like 'and' and 'or' since these make it difficult to distinguish which parts of the hypothesis you are testing. Newton's statement that '*objects attract each other by means of a gravitational force*' is an example of a good hypothesis. It is simple, clear and testable.

Here are two hypotheses which were presented at a junior science fair[1]:

(a) Aphid-infected plants that are exposed to ladybugs will have fewer aphids after a week than aphid-infected plants that are left untreated.
(b) Ladybugs are a good natural pesticide for treating aphid-infected plants.

The first statement (a) is a much better hypothesis than statement (b). It obviously leads to an experiment to test the hypothesis. Statement (b) is a comparative statement, i.e., it raises the question of 'good' in comparison to what? The potential for ladybugs to reduce aphid infection may be an important motivational concept for carrying out the research, but (b) is not the best formulation of a research hypothesis.

It is tempting to write about planning to 'study' a subject or 'investigate' an issue' but unless one can come up with a statement of a research question or hypothesis, in reasonably straightforward language, your proposal is likely to be in trouble.

> *This project seems to be just data gathering without a research hypothesis. There is no research question. It's fine to measure the incidence of heavy metals in the bottom sediments of a river, BUT so what? If you are not going to use the data to answer a question, then it is not research.*

The US National Institutes of Health (NIH) insist on their PIs enumerating a number of specific aims in any proposal submitted. The following example gives an indication of a series of aims that would convince a reviewer that the goals of the proposal are clear. Note also the amount of detail in these statements, rather than vague generalities.

Examples of clearly laid out research questions.[2]

Reviewers appreciate the clarity and specificity of the formulation of a research question.

Specific Aim 1. To establish whether or not modulation of intracellular ubiquitin levels affects Thymidylate synthase (TS) turnover. Transfected cell lines exhibiting altered ubiquitin concentrations will be utilized for this purpose.

Specific Aim 2. To determine if TS mutants with altered intracellular stability are differentially ubiquitinylated. Mutants that have already been produced and partially characterized will be harnessed for this purpose.

Specific Aim 3. To determine if the surface loop of *E. coli* TS is responsible for that enzyme's relative resistance to ligand-mediated stabilization. This is to involve attempts to "humanize" the bacterial enzyme.

Many proposals do not provide a clear description of the research question they are seeking to address. Often the author of an unsuccessful proposal will call the funding agency to ask about the reasons why the proposal was declined. One technique that a program officer may use is to ask the PI to read over the first page of their proposal and state the research question that the proposal was asking. There are usually two responses. Commonly, the PI fumbles around verbally, unable to come up with a crisp research question. This leads the program officer to suggest that more work is needed to clarify the research hypothesis in the proposal. Alternatively, the PI may indeed be able to formulate a simple and succinct answer in which case the program officer may ask why the research question had not been expressed as clearly in the proposal. It is critically important to have a research question clearly formulated and to state it early in your proposal.

I have some difficulty finding a "research plan" in this proposal. Collecting data seems to me to be the plan!!

— Quote from a reviewer

The next example illustrates the lack of clarity about what research question is being asked.

This is an example, based on an actual proposal, of an attempt to formulate a research question but it leaves the reviewer with little or no information about the research hypothesis being tested. The proposal fared badly given the lack of a well-formulated hypothesis.

The first project objective is to collect as much data as possible for as many types of drivers as possible. The second objective will be to subject a limited sample of drivers to simple training in eco-driving, and subsequently evaluate the effect of this training on their driving behavior and carbon footprint. The first objective will be limited by the amount of data loggers that can be purchased as part of the fund. This is currently capped at 50 data loggers due to the budget. In order to increase the sample size to a statistically solid 150 samples, it is proposed that three consecutive measurement campaigns are conducted over the course of 3 months. Due to the relatively smaller sample size, the second objective will be difficult to achieve, hence care must be taken when selecting the subjects who will take part in the training in order to minimize the variables.

In projects in biological sciences and medicine, there is a very definite insistence among reviewers that the research needs to be 'hypothesis driven'. The research topic must be presented as a testable hypothesis that can be answered unambiguously by the research proposed.

> *When asked what makes a good proposal, one of our interviewees replied: Does it have a testable hypothesis? Is it a question that you can answer by experimentation, and that you can actually test your hypothesis?*
>
> **— Quote from a reviewer**

In some other fields of research, this insistence on 'hypothesis driven' research is adhered to less strictly. Nevertheless, the author should always try to be as specific as possible about the question or questions that the proposal addresses. Ideally, these should be expressed in a few simple sentences that make the question being asked very clear and very specific.

The project provides knowledge that has useful outcomes

Another aspect of a proposal that will help convince the reviewer of its merits is if the project has some clear benefit either to other fields (for example, developing a device or technology that will improve medical diagnosis) or if it has some potential practical outcome that might be beneficial to society. Researchers in the medical and biological sciences have an advantage in this regard, since the research in these fields generally benefit health. However, research in many other fields also has the potential to provide societal benefits, even if these are sometimes in the longer term. The example that is often given is the internet, which was initially developed to facilitate the transfer of large quantities of data in high-energy physics experiments, but has since become an indispensable tool of commerce, and part of daily life.

In addition, many countries that support research funding in the sciences and engineering do so with the expectation that these will lead to economic and social benefits for their citizens. This is often the case for countries with developing economies, but is increasingly true in all countries. In such cases, the CFP may contain language that encourages — or sometimes even requires — practical applications. Pointing to such examples will usually be beneficial. Therefore, if you can anticipate practical outcomes of your work, or if you know of companies that have an interest in these outcomes, drawing attention to this will add to the significance of your proposal. There is a more extensive discussion of impact in Chapter Seven.

The research proposal must respond to what the agency wants

While you may believe that you have an important research idea that will move your field forward, if it does not match what the funding agency is looking for your proposal will not be successful. Therefore, you must read the CFP carefully, and perhaps talk with the relevant Program Officer at the agency and make sure that you understand the funder's priorities. If you are seeking funding from a particular agency then you will need to align your goals with theirs. Dr. William Harris, the founding Director General of Science Foundation Ireland, used to call it the golden rule, *viz*, '**He who has the gold makes the rules!**'

For instance, if the agency is looking for proposals in animal cell technology to increase the expression of proteins in animals, there is no point in submitting a proposal for research in diabetes. Or if the CFP states that a 20% match of the total budget from a company is required, and if your proposal does not have that match, you are wasting your time in writing the proposal. The lesson is to read the CFP very carefully and design your proposal to address the priorities of the funding agency.

DO choose an original research topic, especially one that applies to your local environment.

DO try to avoid incremental research.

DO refer to any possible practical outcomes of your research project.

DO read the CFP carefully and make sure that your research project is consistent with the priorities of the agency as given.

DO NOT be too ambitious in your goals. Keep your project within the scope of the resources available.

2.4 Literature Review: Why this is an Important Aspect of Your Research Idea

As mentioned earlier, it is important to establish for the reviewers that you are familiar with the current status of your field. One excellent opportunity to do so is a literature survey. A concise but comprehensive survey can establish that you know the current work, as well as the outstanding problems in the field. Ideally, the reviewers of your proposal will themselves be very familiar with the field so that they can readily judge from your literature review that you know well what you are discussing and are not trying to cover up a lack of knowledge.

The literature survey also provides the **opportunity to place your research idea clearly in context**, to show that it builds on previous work, and to point out the remaining problem that it will solve, and the new understanding that it will bring. It is important to indicate to the reviewers that the research question that you pose is relevant, current and important.

Since research in almost all fields of science is a truly international endeavor, it is critical therefore to be familiar with and to quote relevant papers in a variety of international journals in your literature review. Using only local (in country) references will suggest to the reviewers that you are not really abreast of the current international science and therefore that it is unlikely that your research question will make an important contribution. Fortunately, with most journals now available online, it is much easier to have access to the latest results from a very wide range of international journals.

> *Only local references are cited. There have been many advances in this technology in recent years. I would have expected references from the leading mining journals. There is a very strong theoretical/ modeling network in this area of hydrodynamics that should be referenced.*

A similar problem exists with using references that are not current. Nowadays, most science fields move very rapidly, and situations can change quickly. Therefore, if you do not refer to papers published within the past few years, the reviewers are likely to conclude that your knowledge of the field is not up to date. If there have been no relevant papers published recently, you should state this explicitly in your survey. However, be careful that you are correct since the reviewers may know of any counterexamples that exist.

> *I always look at the references in proposals that I review to be sure they are up to date. Also if I know that the work has already been done, this will kill the proposal. Reading the literature is critical.*

(Another mistake, unfortunately fairly common, is that the first task of the proposal is designated as 'carrying out a review of the literature'. This immediately gives a bad impression to the reviewer, as he or she would expect the authors to carry out a comprehensive literature survey **BEFORE** submitting a proposal, and not after it is funded. Otherwise how would the PI know that the proposed project is original and significant?)

This piece of a literature survey is an excellent example of how to provide such a survey. The references given in the survey [6] through [15] were all to international journals of high quality.

Brief survey of previous research

The use of carbon nanotubes in solar energy applications is a very active and fast-growing research field.[6] For example, carbon nanotubes have been implemented as electron acceptor materials in the photoactive layers of organic photovoltaic devices,[7] or as transparent electrodes for electrical current collection from the surface of thin-film solar cells.[8] Recently, high efficiencies have been reported for polymer/carbon nanotubes hybrid photovoltaic junctions.[9,10] In one case, the high efficiency was attributed to an effective dissociation of electron-hole pairs (excitons) at the nanotube/polymer interface,[9] and in the other to the polymer doping of the nanotubes at the interface leading to a built-in voltage that drives the efficient exciton dissociation.[10] However, the use of carbon nanotubes as the main photovoltaic element in a cell was only demonstrated recently, as mentioned earlier in section I.[5] The later device, as mentioned earlier, is based on the formation of a p-n junction along a single nanotube by electrostatic doping using a pair of split gate electrodes,[11,12] and exposing it to a focused light beam. Very recently, some other groups confirmed experimentally the photovoltaic effect in a single carbon nanotube,[13] few nanotubes connected in parallel,[14] and nanotube films.[15]

DO place your research idea clearly in the context of the current state of the field.

DO indicate why your research idea will make an important contribution.

DO make sure to include relevant papers from international journals in your literature survey.

DO make certain that the references in your literature survey include the most recent appropriate references.

DO NOT *only* refer to local (in country) literature.

DO NOT propose a 'review of the literature' in your work plan.

2.5 Presentation of the Research Idea

In many competitions for funding, large numbers of proposals are submitted and reviewers finish up reviewing quite a few. As a result, reviewers often find that all proposals in a particular area start to look alike. If your proposal looks a lot like the others it could be passed over. Remember this is a **competition for funding** and only a modest fraction of the submitted proposals are successful. Therefore, it is important to try to find some way to make your proposal stand out for the reviewers in a positive way.

One thing that is essential is to state your aim(s) clearly and briefly at the very beginning of the proposal, in the form of a testable hypothesis if at all possible. Reviewers will be looking for this, and if they do not find it early you will have a hard time getting their attention as they read on.

The importance of clarity cannot be overstated. Even though you may think that you have stated your research question clearly, you need to be sure that a reader, who may not be intimately familiar with your area of research, will also find it clear and unambiguous. One very simple technique is to **read the statement of your research idea aloud**. Often this will help you to realize that the statement as written is not as transparent as the one you have in your head. Another good way to test for clarity is to have a friend or colleague read the statement to see if they can understand it easily. We will return to the importance of other readers later in Chapter Four.

There are many ways to turn off a reviewer. One of the easiest (and one of the most annoying for reviewers!) is to have incorrect spelling and poor grammar. The reviewers we interviewed were unanimous in finding this annoying. If reviewers find poor spelling and grammar on the first page, they will often conclude that your approach to research is also careless, and they will be less inclined to read the remainder of your proposal with a positive attitude. Therefore, you should use all the tools available to make sure your use of language is correct, especially the statement of your research aim(s).

2.6 How Important is it to Present Preliminary Data as Part of the Proposal?

The use of preliminary data in proposals varies somewhat between scientific fields. In the general area of biological and medical sciences, preliminary data, which gives support to the proposed hypothesis, is becoming more and more important. Most reviewers in these areas will mark down a proposal severely if there is no preliminary data included with the proposal. This is usually not a problem for established researchers who have ongoing funding, but can be an issue for a new researcher who does not. In such cases, the usual source of funding is from internal funds within the institution, but, of course, these are usually quite limited. In some cases in biological and medical science, it may be better to delay submission of a proposal until preliminary results can be obtained.

In other fields of science and engineering, the inclusion of preliminary data is not so critical. However, if data are available, even if unpublished, it is usually useful to include such data in the proposal. **The more controversial the research aim of the proposal, the more useful it is to supply supporting data.** Similarly, if your research goal depends on a new technique or new piece of equipment, presenting even a small amount of preliminary data will help convince reviewers that you can achieve the goal.

2.7 Multidisciplinary Research Questions

Many research questions, particularly those that relate to real world situations and are, therefore, of a more applied nature, often involve more than one academic discipline. This poses special problems in developing a proposal, since it often means involving collaborators with expertise in different disciplinary areas to address the research question.

Reviewers will usually need to be convinced that the various collaborators are really engaged in the project, since their participation is essential for success. Letters from collaborators attached to the proposal is one way to deal with this issue. Or simply stating exactly how they will participate helps. If some of your collaborators have already participated with you on other proposals and have even published papers with you, this provides some evidence that you have a good working relationship. You need to point this out. In any case, you will need to pay some attention to making sure that the reviewers are certain that all your collaborators from various disciplines are willing to do their part to make the project a success.

> *In this proposal, the reviewer noted that the PI and a postdoc were in the budget to spend some time in a laboratory in Japan. This made sense to him, because this is where scientists from XXXX can learn forefront techniques and bring those back home. It also gives them exposure to new ideas.*

How do you find appropriate collaborators? Often the problem itself will suggest not only the area, but even specific people who are interested in the same problem. There are, of course, advantages in working with collaborators in the same institution. You can meet easily face to face on a regular basis to report progress, discuss issues and solve problems. However, the internet has now made it possible to communicate readily across considerable distances that would have been difficult or impossible even a few years ago so this broadens the scope of possible collaborators considerably.

The issues around working with collaborators will be explored in more detail in Chapter Six.

2.8 Should the PI Meet with Agency Staff before Preparing a Proposal?

Every funding agency, whether governmental or private, will have its own priorities for funding research. While this information is generally available either in print or on the internet, we believe that it is also very useful to discuss particular ideas with program staff of the agency. This will often help give the researcher a clearer understanding of the current priorities in the agency, and could suggest a particular research idea that fits well with the priorities of the agency.

Meeting with agency staff and discussing your research also gives them an appreciation of your motivation and enthusiasm for your work. While most agency staff try very hard to be objective and professional in their decision making, it cannot hurt if they can put a face and a voice to a particular name. It is surprising how few researchers contact Program Officers in person. More experienced researchers often do this.

Agency staff are often helpful in discussing an unsuccessful proposal that the applicant intends to resubmit. They can help interpret the earlier reviews and point out directions that may prove more fruitful. In cases where a proposal has been discussed in a panel review process, the staff may be able to provide additional information that was not written down in the reviews. These insights can be helpful in preparing a resubmission.

2.9 Summary

- A good research idea is critical to a good proposal. Good ideas move the field forward and therefore it is important to be very familiar with the current status of the field.
- Keeping at the forefront of your field means reading the published literature, attending seminars and communicating with colleagues both locally, within your institution, and also at national and international conferences.
- Don't try to solve all the problems in your field in a single proposal. Be aware of the limitations of resources available in a particular competition. Make sure your research aim is specific and addresses a particular scientific question.

- Express your research aim as a testable hypothesis as far as possible.
- Emphasize the importance of your project early in the proposal to catch the attention of reviewers. The importance can be because of its potential impact on the science or the potential practical outcomes.
- Provide preliminary data, particularly if your field is biology or medicine. In other cases provide data if available. This is more important if your project is controversial and the basic concepts might be questioned by reviewers.
- Take the opportunity to visit the staff of the funding agency to see if they can give you additional insights into what the agency is looking for currently.

DO present a crisp, well-formulated research question early in your proposal.

DO make sure that you indicate, using up-to-date international literature, how your proposal builds on what is already known about the topic.

DO emphasize the significance of your project by pointing out how your research question will contribute to the knowledge base and/or to benefit of the country supporting the research.

DO make sure that you provide evidence of collaboration if this is needed to address a multidisciplinary research question.

DO NOT only refer to local literature in reviewing the field.

DO NOT simply state that you are going to 'study' or 'investigate' some phenomenon as your research goal.

DO NOT assume that the reviewers will be familiar with the specific field of the proposal.

DO NOT annoy the reviewers by having poor grammar or poor spelling especially in expressing your research idea.

Chapter Three

The Review Process

Democracy is the worst form of government except for all those others that have been tried.

— Winston Churchill

Well, now you have an idea that you believe is exciting and will make a difference to your research field. Before you start actually drafting your grant proposal, it's worth thinking about who will be reading your proposal and evaluating it.

In preparing any written work, it is important to know your audience. This is particularly true in the case of preparing a proposal for review. In this chapter, we shall discuss the review process and provide some insights about the people who will read your proposal, and the different mechanisms involved in the review.

Galileo's 'heliocentricity' grant proposal is assigned to the wrong review panel. *Again.*

In addition to budgets, one of the main issues that funding agencies must deal with is how they review the proposals received in response to a Call for Proposals (CFP). Almost always, there are not enough resources to fund all good proposals so a competition is used to select the very best proposals for funding. The most common process for reviewing proposals is peer review, where the review is carried out by experienced people who are active researchers in the general area of the proposal. In some cases, the staff members of the agency handling the proposals are also professionals in the various areas of interest and can offer opinions on the proposals. However, in most cases, peer reviewers are from outside the agency.

Peer review is not a perfect process for establishing the merit of a proposal, but it is the most common process used by experienced research funding agencies to evaluate proposals. It is, like 'democracy' referred to in the quote above, the best review process available.

In countries with a large number of active scientists in any given field, the majority of reviewers can be scientists working in the same country. Precautions need to be taken so that there is no obvious bias, either positive or negative, in the review process. For example, collaborators or previous research advisors are excluded, and a notification of any conflict of interest is usually required from reviewers. Normally this is not a major issue. Even in countries with large numbers of researchers in particular fields, it is still sometimes the case that international reviewers will be included in the mix in order to present an even broader point of view.

However, in countries with fewer scientists in any given area, the possibility of a conflict of interest (or the appearance of such a conflict) becomes much greater if local scientists are used as reviewers. Therefore, in such cases **funding agencies will often enlist scientists from outside the country as reviewers**. This adds to the cost of reviewing but **provides a necessary objectivity to the process**. Using international reviewers also provides a useful function in helping the funding agency (and applicants themselves) benchmark the proposals against international competition.

For less mature or less established research systems, it can also provide a **valuable source of advice from researchers in other countries** about professional research practices, and what would make the research more competitive.

In cases where foreign scientists are used as reviewers, there is the problem of a common language. English has now become an almost universal scientific language, and so most proposals that are reviewed by reviewers from outside the country are written (and reviewed) in English. This places an additional burden on authors of the proposals from non-English speaking backgrounds, who must work in a foreign language. This can be an added burden on the reviewers who, in many cases, have to make sense of proposals that are written in poor English. **Nevertheless, the advantages of using expert reviewers from outside the country far outweigh the additional burdens involved**.

3.1 Initial Screening of Proposals

In most cases, the first step in the review process is for the agency to check that the rules spelled out in the CFP are being followed. This includes checking the eligibility of the applicants, the time of submission, the page length both by section and overall, plus any other special requirements (such as 'matching funds' from industry). This process is normally carried out by the staff of the agency since no special expertise is needed — simply a careful reading of the CFP. Normally any violation of the rules, even the most trivial, will cause the proposal to be returned without review. This process should make the applicant very aware of two truths about grant funding:

- Your proposal is **entered in a competition** and any reason you provide to allow it to be penalized will harm your proposal and may even cause it to be excluded.
- **You should read the CFP very carefully and follow the guidelines strictly and accurately**.

One of the most common mistakes that applicants make, especially if the process is new to them, **is to submit their proposal late**. This is one of the most frequent reasons why proposals are returned without review. Once the deadline has passed, even by as little as a minute or two, it becomes impossible to set another time as the deadline. If 2 minutes late is allowed, why not 5 minutes or 30 minutes, and so on? Therefore most agencies will be strict and unforgiving about late submissions, because this is the only way to be fair to all applicants. As more proposals are submitted electronically, enforcing the deadline is even easier. The lesson is to make sure that your application is submitted in plenty of time, and allows for any glitches in the submission process at your institution.

Similar considerations apply to limits on page length, which is sometimes applied to the proposal as a whole, and sometimes to individual sections, like the Curriculum Vitae. In some cases of excess length, rather than returning the proposal to the applicant, the agency will simply truncate the document to fit within the CFP guidelines. This will mean that your proposal appears incomplete, and will make a poor impression on any subsequent reviewers.

> *"The methods section ended in the middle of a sentence and therefore was clearly incomplete. This suggested to me a lack of care in the preparation of the proposal"*
>
> **— Quote from a reviewer**

3.2 Selection of Reviewers

Reviewers are selected by the staff of the agency. The goal is to select reviewers who are experts in the subject matter of the proposal and who can provide an informed opinion about the value and likelihood of success of the proposal. Reviewers are expected to be objective and unbiased and to base their opinions on the criteria described in the CFP.

It is also generally the case that a reviewer will rate more than one proposal in any particular competition, so that a certain amount of normalization of a reviewer's assessments can be obtained. Some reviewers are more or less critical of others' work, so having reviewers read a number of proposals allows the staff of the agency to weight any individual reviewer's scores, in order to be fair to all applicants. Because a range of reviewers, with varying degrees of specialization in your area, may be reading your proposal, you need to be very careful to make your proposal understandable to reviewers who may not be as intimately familiar with your proposal area as you might prefer.

There are agencies where the authors of a proposal are asked to suggest reviewers, but this often leads to a biased review and the practice is not widespread. If you are given this option, there are some principles you should follow when making suggestions. (See box)

DO NOT nominate people with whom you have published or collaborated.

DO use the opportunity to identify the types of people that are most appropriate for your proposal (for example, what is the best mix of disciplines/ sub-disciplines?)

DO NOT nominate famous researchers in your area who are likely to be too busy to review your proposal. (You are wasting a nomination.)

DO nominate researchers whom you believe would be in a position to give substantial constructive advice that could strengthen your project.

In some cases, a funding agency will permit proposals to be submitted at any time, with the proposals reviewed as they are received. The agency staff then make a funding decision after receiving enough useful reviews. This has been the common practice at the US National Science Foundation, and also in the early days of Science Foundation Ireland. However, it is now more common for competitions to be run with a deadline for proposals, so the proposals compete for funding at the same time. This is certainly a more common mode of operation in small countries and newer research funding agencies, since it is more efficient and simplifies the budgeting.

The breadth of topics in any particular competition also determines, to some extent, the choice of reviewers. If the choice of topics is very large but the number of proposals submitted is not, it may be necessary to use reviewers who are less expert in the particular topic being proposed. This situation will usually arise in a small country running a competition with a specific deadline and a very broad scope.

One recent example was a competition where 157 proposals were received in the broad topic area of 'Physical Sciences and Engineering'. These proposals ranged widely in topic from **general relativity** to the **wear of machine tools with repeated use**. Finding the 50–80 reviewers who could provide the 471 reviews needed (at least three reviews per proposal) proved to be a challenging task, and only allowed a comparatively small number (5–10) of proposals to be read by any one reviewer. This illustrates how difficult it is to find reviewers who are truly experts in all the fields in which they are being asked to review, when proposals covering a very wide area of science are being reviewed at the same time.

The lesson for proposal writers is that it is very important to write your proposal so that it is understandable for a reader/reviewer who may not be an expert in your very specific area. **This means no jargon, few acronyms and — especially — that you write simply and clearly**. There is always some degree of balance required between providing enough detail to make your proposal credible, and at the same time NOT making it difficult to understand.

As mentioned earlier, particularly for small countries with a limited number of local scientists, the use of reviewers from outside the country provides a more independent and unbiased review of the proposals. An additional advantage for those countries still developing their scientific infrastructure is that international reviewers provide a good measure of the progress of the local scientific community compared to international standards. The reviews, which are normally shared with the authors of the proposals, then provide useful advice to the applicants, even the unsuccessful applicants, on how to improve their proposals in the future. International reviewers also create awareness among the international science community of the efforts within the country to build a professional high quality research system.

3.3 Postal Reviews

By far the most common type of review is a postal review. The reviewer is asked to read a proposal and write a brief review. Almost always the reviewer is also asked to provide a numerical or alphabetical rating for the proposal. Sometimes the quantitative scoring is done section by section against specific criteria, or sometimes only an overall summary score is required. In some cases, postal reviews are supplemented with a panel review in order to compare proposals against one another and to prioritize them if necessary.

While it is tempting to use only the summary ratings in comparing proposals, agency staff are strongly encouraged to read the text of the reviews carefully before making a recommendation as regards funding. In some cases a reviewer may have found a 'fatal flaw' in a proposal that other reviewers may have missed, and which would prevent, or at least argue strongly against, the proposal being funded. For example, the reviewer may be aware that the work proposed has already been published, or may know that the technique being proposed will not work in this situation. In such cases, the average of the quantitative ratings may be largely irrelevant.

In most cases, a minimum of three reviews is required. However in complex proposals, many more reviews may be used, sometimes as many as a dozen. The quality of the advice improves with more reviews but, of course, this entails more expense. There is a trade-off between cost and the quality of the final decision.

In a few cases, agencies conducting reviews of proposals allow a response to the reviewers' comments before a final decision on whether to fund a proposal or not. This is discussed in more detail in Chapter Ten, but generally the process involves a written response from the applicant. The US National Science Foundation does not have a reviewer response option, for instance, but it is more common in Europe and Australia. The response may be restricted to pointing out any factual errors in the reviews, or may be more generally responsive to the whole review. The authors of proposals generally like to have the option of responding to reviews, although it is not clear that these responses make much significant difference to the quality of the process. This step also adds time to the review process, when in most cases, funding agencies are trying to reduce the time required to complete a review process.

3.4 Panel Reviews

Another method that often improves the quality of the review process is the use of a panel of reviewers who meet together and evaluate a set of proposals in a particular area. The panelists can be those serving as postal reviewers or can evaluate postal reviews submitted by other reviewers. The advantage of having the panelists serve as postal reviewers is that they can defend their review in the presence of peers, particularly helpful where there is disagreement between reviewers. The other advantage, even more practical, is that it ensures that all reviews are received by the time of the panel meeting.

A panel review is particularly valuable in cases where the staff members of the agency are not professional scientists or at least are not very familiar with the scientific area of the proposals under review. In such cases, a panel of experts who meet and rate all the proposals in a particular area can be very helpful to the staff.

"Is it just me, or are these review panels getting a *lot* tougher?"

Of course, chairing the panel must be handled carefully to gain the most benefit. Agency staff will sometimes serve as the chair of the panel or the chair may be selected from among the panel members. Even in cases where the chair of the panel is one of the scientific reviewers, an agency staff member should be present in the meeting to help answer questions about the competition and to help keep the discussion on track. The chair must also be careful to ensure that no panel member dominates the discussion or the decisions. An outspoken member with strong opinions can unfairly influence the rest of the panel, so the chair needs to be careful to solicit views from all the members and to avoid any undue influence by a single panelist.

When using panel reviews, it is probably even more important to use international reviewers if the number of active scientists in a particular area in the country is small and there is even an appearance of a possible conflict of interest. In the case of Science Foundation Ireland, panel reviews were used to evaluate the Research Frontiers Program for a number of years. At first, the staff decided that having one Irish scientist on a panel of typically about a dozen members would be educational for the Irish science community. This was a controversial issue for local scientists. They believed, incorrectly, that the local Irish scientist exerted an undue influence on the panel and would support his or her friends' proposals at the expense of others. As a result the practice was discontinued so that all panel members were international scientists. This experience reinforced the view that the use of international reviewers is very important when the size of the local pool of scientists in any given science discipline is rather small.

There is one other advantage in using international reviewers in cases where the country organizing the review is trying to build a more internationally visible scientific and engineering community. International reviewers, having read proposals and discussed them with colleagues will be in an excellent position to obtain an overview of the local science and engineering scene. In contrast with simply serving as a postal reviewer, a panel member gets exposed to a greater number and range of proposals, and is better introduced to the funding agency and its budgets. Panelists at Science Foundation Ireland, for example, were often favorably impressed by the quality of the proposals they reviewed. This undoubtedly improved the image of Irish science as these reviewers returned to their home countries and discussed their impressions with their colleagues.

When your proposal is reviewed by a panel, you will normally be asked to nominate which panel should do so. This decision is one you should consider carefully, especially if your proposal has an interdisciplinary aspect. For instance, if your work has more obvious near term applications, an engineering panel may look upon it more favorably than a pure physics panel.

In some cases where many proposals are anticipated for a particular competition, the funding agency will ask for a brief pre-proposal to be submitted for initial screening. This is a means of eliminating projects that are deemed not competitive without taking up large amounts of time and effort on the part of both the applicant and the reviewers. Of course such a 2 stage process also has disadvantages, one being that it takes considerably longer to reach a funding decision.

One common practice for ranking proposals by panel review is particularly worth being aware of when preparing your grant proposal. Many panels as they discuss individual proposals, will initially assign each application to one of three recommendation categories: (1) 'Definitely fund', (2) 'maybe fund', (3) 'do not fund'. Typically only a small fraction of proposals — say about 5%, — fall into the 'definitely fund' category after a first discussion of the proposals, with perhaps 30–40% in the 'maybe fund', and the remainder 'do not fund'. Normally the funding agency will also have given some indication of how many proposals (or the amount of money available to fund proposals) they are likely to be able to fund, and this may only allow for funding about 10–15% of all the proposals submitted. **In this context, the key challenge for many good proposals is to make sure they survive the process of attrition** as the review panel goes through the proposals multiple times looking for reasons to assign proposals to the 'do not fund' category. Under these circumstances certain factors which do not necessarily make your application a 'bad proposal' will make your application **less competitive** relative to others. At this stage in the process, issues such as the proposal having unrealistic timescales, an inadequate level of detail, uncertainty about the role of collaborators, or even the clarity or quality of the writing might be enough for your proposal to be moved from the 'maybe fund' category to 'do not fund'.

As discussed later in Chapter Ten, it is important not to forget that you are part of a **competition**. Many inexperienced grant applicants who are declined for funding, even though their proposal has received quite positive reviews, can be upset by the decision. *If the reviewers have said my idea and CV are good*, they say, *then why haven't I been funded?* The answer is that although the proposal may be good, the case for funding it may not have been communicated as well as some of the other competing proposals. And, unfortunately, sometimes the reasons a proposal is assigned to 'do not fund' are all too avoidable. A little more care taken with, for example, the quality of the writing, the level of methodological detail, or the case for the relevance and impact could have made all the difference between success and failure.

3.5 Site Reviews

For more complex and expensive proposals, in particular for large research center grants, a site visit is often used as part of the review process. Such proposals usually involve a number of co-principal investigators and a large budget, normally in the range of millions of US dollars. In such cases, the agency will often invite a small panel of 3–5 reviewers to visit the site where the research is to be carried out. These reviewers will have access to previous postal reviews of the proposal and sometimes even the response to these reviews from the authors. This provides the opportunity for the panel of reviewers to hear from the principal investigators (PIs) about the proposal, including the opportunity to question them about any of the concerns raised by prior reviews. It is also an opportunity for the reviewers to 'interview' the lead PI or center director and try to evaluate their personnel skills since their leadership is critical to holding the large team together and making the collective research effort a success. Because of the scope, complexity and critical mass of these large grants, site visits also give the panel the opportunity to question the PIs about their plans for the management and governance of these major research endeavors.

A site visit is also an opportunity to gauge the enthusiasm of other co-PIs for the collective endeavor — of making the whole greater than the sum of the parts — or whether they are likely to just do their own individual research as a 'silo' within the larger grant.

There are additional advantages of a site visit. The panel will usually hear from the administration of the research facility and can therefore gauge how much support can be expected from the management. Similarly, if collaborators are involved, the level of their commitment can be assessed, and the contributions that can be reasonably expected from any industrial partners evaluated. For example, if the Vice President for Research from Proctor & Gamble attends a significant part of the site visit, this indicates a real commitment of the company to the project being reviewed.

A site visit also provides the opportunity for the review panel to see the facility and infrastructure at the site where the proposed research is to be carried out. This can give a good idea about the preparedness of the team and also give guidance about the equipment requested in the budget, and whether it is really needed or not.

Finally, the review panel will have the opportunity to meet with younger researchers and even students who may be involved in the project and this can give a useful guide to the enthusiasm and management style of the PIs. All of these factors are very useful in assessing the likelihood of success of a project.

A so-called '**reverse site visit**' is used sometimes to provide some of the advantages of a full site visit but at decreased cost. In a reverse site visit, the PIs are invited to present before a panel, but the meeting usually takes place at the agency and not at the location where the research will be carried out. Again, this is an opportunity for the reviewers to 'interview' the lead PIs to explore their ideas and strategy, as well as their leadership qualities, which can be critical to making the research effort a success. This also gives the panel the opportunity to question the applicants, including some of the collaborators, which can be very useful. However, first hand information on the facility where the research is to be carried out is not available to the panel. Nor do they normally have the chance to meet with representatives of the administration of the home institution and to assess their enthusiasm and support for the project.

A reverse site visit is often used when there are multiple competing proposals, not all of which can be funded. The same panel is used for all the proposals and the review of multiple proposals can be carried out over a few day period.

Even in cases where an award is to be made to only a single PI, some countries, such as Japan, will sometimes require an interview as part of the review process. It is certainly true — as anyone who has participated in such interviews will agree — you learn a great deal from hearing and watching a PI respond to questions.

3.6 Summary

There are a number of different approaches used for reviewing proposals:

- **Postal reviews**, where a reviewer submits a written evaluation of the submitted proposal based on the criteria in the CFP.

- **Panel reviews**, where a panel of reviewers meets, discusses and evaluates a group of proposals. These same panelists may have already written postal reviews or have access to postal reviews from other reviewers.
- **Site visits**, where a small panel of reviewers meets with the applicants to make an assessment of proposals. In some cases, with more than one proposal to consider, a **reverse site visit** is carried out where the PIs of the various proposals meet at a common site.

In countries that have only a comparatively small number of scientists in any specific science field, international reviewers are often used. **This has the advantage of objectivity and the lack of the appearance of any conflict of interest**. However, it is more expensive and often requires the proposals to be written in a foreign language, usually English. The other advantage of using international reviewers is that it provides a window into the science of the home country for the international community and an opportunity to showcase the strengths of the in-country scientific community.

In submitting a proposal for review:

- **DO** read the CFP carefully and follow all the requirements and guidelines.
- **DO** submit your proposal before the deadline. This means that you need to allow time for your proposal to be processed in your home institution.
- **DO NOT** exceed any page limits either in individual sections or the proposal as a whole.
- **DO** make sure that your proposal is written to be understood by a reviewer who may not be an expert in your very narrow field.
- **DO** bear in mind that you are in a competition. Your task is to convince the reviewers that your proposal merits funding in comparison to the others. All sections offer an opportunity to compete. Completing any section inadequately may cost you.
- If your proposal in a multidisciplinary domain, **DO** write so that your proposal can be easily understood by researchers from all relevant fields.
- **DO NOT** use jargon, unexplained acronyms, etc.

Chapter Four

Drafting the Proposal

If people knew how hard I had to work to gain my mastery, it would not seem so wonderful at all.

— Michelangelo Buonarroti

Genius is 1% inspiration and 99% perspiration.

— Thomas Edison

Ok, so now you have a research project in mind and have some idea of the kind of reviewers who will be reading your proposal. It is now time to turn to the mechanics of actually making your proposal excellent.

There are many proposals with a good, even superb, idea but which fail because of poor presentation. There are many ways in which a proposal can be presented poorly; lack of clarity in the writing, using too much jargon, **or simply not following the directions in the Call for Proposals** (CFP). You need to make sure that not only is the grammar correct but also that the logic, coherence and structure of your argument are all sound.

Another important piece of advice, which cannot be stressed enough, is to **give yourself plenty of time to prepare the proposal**. In most cases there is a deadline announced for a particular competition. While it may be tempting to delay starting to prepare your proposal in order to get more initial data or to read more background information about the subject, be warned that everything takes longer than anticipated. Therefore, make sure you have allotted plenty of time to prepare the proposal, to provide time for review by others and to check it again carefully before submission.

Having a knowledgeable colleague read the proposal when it is close to a final draft is also very helpful. Another set of eyes can often help you clarify your message and catch errors or other problems with the text.

In this chapter, we will outline good practices for reading the CFP, offer advice on effective approaches to writing key sections, including the abstract, introduction, literature review, methodology, and the CVs of the leading researcher and collaborators if required. We will also provide real examples of some of these sections taken from successful proposals.

4.1 Read the CFP Document Carefully

Reading the CFP document carefully may seem like obvious advice but, unfortunately, it is not always followed. The funding agency has undoubtedly put a good deal of thought into this document to try to elicit the kinds of proposals that best suit the goals of this particular competition. Agencies will want to fund the science that will make a difference but there may be specific additional goals, for example, encouraging cooperation between different fields of science or between academia and industry. These should be clearly mentioned in the CFP, and should be explicitly addressed in your proposal.

There will also be a number of criteria listed in the CFP upon which the reviewers will be basing their judgment. You **will need to make sure that your proposal responds to each of these criteria explicitly** so that the reviewers will find it easy to see that your proposal does indeed match the criteria required for this specific competition.

> **One of our interviewees, who is a very experienced and successful grant-writer, uses the following technique:**
>
> When writing a proposal, he first goes to the CFP and pulls out the things the funding agency is looking for and copies them into the early outline of his own grant document. He colors them in **RED**. If there is anything else in the grant announcement that is important, and there may be very specific things, he also puts those in his grant document in **RED**, e.g., relationship with other agencies, cost sharing etc. They all go into the initial proposal document in **RED**. The **RED** coloring is a signal that this point needs to be addressed. When the point is covered during the writing of the proposal, the statements in **RED** are deleted so that when the proposal is completed, there should be **NO RED** left. This ensures that the proposal has addressed all the items that the agency is looking for and that are listed in the CFP.

Normally in a CFP, there will a **format and font specified, along with page limits or word limits** both for the overall document and/or for specific subsections. It is extremely **important to adhere to these limits** since failure to do so may result in your proposal being declined without even being reviewed. It would be the height of folly to spend many days or weeks in preparing a proposal to find it returned because the page limit was exceeded.

If the proposal is not immediately rejected, the agency can simply administratively remove any additional pages, even if your text is in the middle of a sentence, once the page limit is reached. This will look very bad to the reviewers and will jeopardize the success of your proposal. The reason that agencies have to be so rigid in their treatment of excess pages or words is that this is the only way to be fair to all applicants. If the agency accepts excess pages from one applicant, they must do so for all applicants, and the page limit becomes meaningless. Even if the formatting and the page limits are only specified as guidelines, you would be wise to stay fairly close to these guidelines otherwise you may annoy the reviewers. Remember that reviewers are busy people with severe time constraints. They normally will have to read a number of proposals and will not welcome trying to read tiny print or excessively long text.

Space limits can be a particular issue when dealing with the applicant's resumé or curriculum vitae (CV). Most scientists have a standard CV that is kept on file and can be pulled up easily and quickly. It is tempting therefore to simply drop this standard CV into the proposal. However, the CFP may require either a specific format or have special requirements, such as listing five of the author's publications most relevant to the work proposed. You must make sure to **follow these specifications and not simply take the first page or two from your standard CV**. Even worse is to ignore the requirements and provide the complete CV if that is not called for. Again the reviewers will be annoyed by having to sift through many pages of material and at best will simply ignore it.

The CV you present with your proposal should be modified it to highlight factors relevant to the specific competition. For example, a track-record of having already worked with your collaborators; or an indicator of time spent in industry if industrial collaboration or application is one of the goals, can be very helpful.

In some cases, but not always, the CVs of collaborators or of co-principal investigators will be requested, or even required. **Including them can be particularly useful if you want to highlight key expertise or experience that they bring to the project**. You should make certain that you comply with whatever is required. You may think that adding the CVs of collaborators is useful even if not required but, by doing so, you burden the reviewers and most will not appreciate this. If the CFP is not clear in this matter, check with the agency to see if collaborators' CVs are either required or if they should not be part of the proposal. **Having a proposal that is unnecessarily bulky is not how you want your proposal to stand out**.

Another requirement where the issue of fairness to all applicants arises is the deadline for proposals to be submitted. As discussed in Chapter Three, enforcing a strict deadline is generally the only way to be fair to all applicants and most agencies follow such a policy. Increasingly, deadlines are enforced by the online systems through which grant proposals are electronically submitted. It may seem like a trivial issue to submit a proposal a few minutes late but the problem that then arises is where to draw the line. If a proposal can be submitted five minutes late, why not 10 minutes late and then why not an hour late and so on. **There is no other clear line that can be drawn, except the original deadline**. The lesson, therefore, for the principal investigator (PI) of a proposal is to make sure that you have sufficient time to submit your proposal comfortably before the deadline. In most institutions, all proposals are required to be checked and signed off by the administration. You need to allow sufficient time to cover all these steps, plus any contingencies that might arise.

The only exception would be if there is some problem with the processing at the agency. In that case, the agency should make allowances: if, for example, their software has some glitch or if there is an overload of their system, the deadline should be revised. Problems with the applicant's own systems or that of her or his home institution will not usually be considered as an excuse. We have probably all heard of cases where a proposal was prepared after months of work but then due to an oversight it sat on someone's desk, and did not get submitted until just after the deadline had passed and therefore was not even processed by the agency. Take care to avoid this happening to you.

Another section of the proposal in which the requirements of the CFP are usually particularly stringent is the budget section. The budget is so important that we devote a whole chapter to this (See Chapter Nine). The main point is that you will need to follow the instructions included in the CFP as regards the budget and how much detail is required.

DO read the CFP very carefully.

DO address all the criteria listed in the CFP as explicitly as you can.

DO follow all the page limits, formatting and font requirements in the CFP, including those for the CV and the budget.

DO NOT submit your proposal after the deadline, even by a minute.

4.2 Cover All the Sections of the CFP

Often a CFP will break down the proposal into sections. As mentioned above, these sections might have individual space limitations and there is almost always a total page limit. Typical sections of a proposal might be:

- Abstract.
- Introduction.
- Statement of the problem and its context, including the relevance of the project and its potential impact.

- The description of the research methods used: this may also require a work plan including timelines, milestones, and outputs.
- CV of the applicant with relevant publications.
- Budget, which will likely include a justification of the resources requested.
- Letters of support from partners.
- A statement or at least a signature from the host institution agreeing to the conditions of the competition.

Let us examine each of these sections briefly.

Abstract

The abstract provides a short summary of the project and will normally be limited in size. This is one of the most important sections of the proposal since it is almost always the one that a reviewer will read first. It is important that the abstract catches the reviewer's attention by stating clearly, **WHAT** the proposal is intended to do, **WHY** this is important and, briefly, **HOW** it will be carried out. This is the place to highlight the 'wow' factor in your proposal — the issue that makes it truly unique and makes it 'stand out from the crowd'.

If appropriate in the space allowed, it is also useful to refer to any previous results that are relevant, and any previous experiences of the author that will convince the reviewer that he or she has the expertise to carry out the work proposed successfully. The details of the methodology can usually be spelled out in detail later but, if there are unusual aspects of the methods, or, if the project requires special equipment or other collaborators, these could also be mentioned in the abstract.

In the following block, we give an example of an abstract from the field of Oceanography provided by Professor C. Benitez-Nelson from the University of South Carolina. Note that the abstract clearly outlines the project, states why the work is important, and briefly discusses how it will be carried out. There is also a mention of preliminary data, and the goals of the project are stated clearly and concisely. Note in particular the specificity in the project and the emphasis on the broad scientific impact.

PERSISTENCE AND FATE OF DOMOIC ACID IN THE SANTA BARBARA BASIN

Toxic blooms of a variety of algal species (harmful algal blooms (HABs)) have been documented throughout the world's coastal oceans, ultimately impacting shellfish, finfish, marine mammals and birds over large areas. Several species within the genus of *Pseudonitzschia*, a group of marine diatoms that produce the neuro-toxin domoic acid (DA), have been identified as common members of algal assemblages along the coast of California. Key questions in HAB research include not only what causes toxic *Pseudo-nitzschia* spp. to bloom, but what happens to that bloom after its demise. Causative factors ranging from coastal eutrophication to increased upwelling to resuspension of seed populations from sediments have all been hypothesized, but remain enigmatic, mainly due to a paucity of integrative data. The fate of DA producing *Pseudo-nitzschia* blooms remains even more elusive, and yet there is increasing evidence of substantial DA concentrations in benthopelagic feeders and benthic organisms both along the coast and offshore.

Why

This proposal seeks to build upon intriguing and exciting preliminary data that suggests that sinking particles are a major vertical transport mechanism of DA from surface waters to sediments, with DA fluxes exceeding 50,000 ng DA m-2 d-1 at depths in excess of 500 m. We therefore hypothesize that DA is rapidly transported to sediments and likely persists on greater than seasonal timescales, well after the demise of a *Pseudo-nitzschia* bloom.

Hypothesis

The goal of this proposal is to create a regional observation and modeling program focused on the Santa Barbara Basin (SBB) that specifically: (1) Examines the temporal relationship between Pseudo-nitzschia blooms, DA toxicity, and the vertical transport efficiency of Pseudo-nitzschia and DA to the benthos, (2) Investigates the incorporation of DA into the sediments, and (3) establishes a historical record of DA toxicity spanning the past decade, and potentially, the last century.

What

To achieve this goal, we will use a combination of monthly survey cruises of water column biogeochemistry, two continuously moored sediment traps (150 and 550 m), regional surface and

(Continued)

(Continued)

down core sediments records, satellite imagery and develop a numerical model that examines the surface timing of toxic *Pseudo-nitzschia* blooms and the vertical export of DA to the seafloor. This transformative research will fundamentally change current views on the persistence of DA toxicity in marine systems and will provide the groundwork for similar measurements of other harmful toxins. The proposed research leverages sampling opportunities with a number of ongoing programs off coastal California, such as Plumes and Blooms and the Santa Barbara Coastal LTER Project (water column), the Marine Sediments Laboratory at USC (archived and future sediment trap collections), and the Southern California Coastal Water Research Project Bight 2008 campaign (surface and down core sediment collections).

How

The abstract does not have to be very long and in some competitions there are tight constraints on the length of an abstract. We provide another example of an abstract from the physical sciences that conveys the essence of the proposal and its importance in only 118 words.

Abstract

The ability to control the transport of small molecules through very small pores in membranes is important for a variety of applications such as the development of drug delivery systems and separation techniques. An improved understanding of the fundamental chemistry that controls the transport of small molecules would have broad biomedical, biotechnology, and environmental implications. This work focuses on regulating the transport of small molecules through gated nanopores and creating nanocapsules capable of rapid uptake and release of small molecules. The transport will be regulated by controlling the chemical environment of nanopores and by reversible binding of plugs or lids. This interdisciplinary project integrates research and training with outreach activities aimed at underrepresented groups in science and engineering.

From Professor E. Pinkhassik, University of Connecticut

Your abstract will be used by the staff of the funding agency to assist them in allocating your proposal to a particular review panel or in deciding to whom to send your proposals for review. It is important therefore that the abstract accurately reflects the key elements of your project and gives enough information for a reviewer to assess whether the proposal is appropriate to their expertise.

In your abstract,

- **DO** make sure you answer WHAT? WHY? and HOW?
- **DO** present the unique aspects of your proposal.
- **DO** mention any critical equipment or important collaborators.

Introduction

In the introduction, the summary of the project presented in the abstract can be elaborated upon since more space will be available. You should again make sure to state clearly **WHAT** your proposal is about, **WHY** it is important and **HOW** (briefly, since the details will be covered in subsequent sections) you plan to carry it out. It is very important to explain the importance of the project, and how it relates to the goals of the competition. For example, if one of the goals of this particular competition is to promote industry–academic collaboration, the introduction is the place to mention your industrial partners or any special activities, like internships, that may be appropriate. Remember, the introduction is the first part of the proposal proper that the reviewers will read. **You must capture their attention and give them a good first impression of your proposal**.

By way of illustration, in the following block we show a part of the Introduction from the same proposal from which we presented the abstract earlier. Note the expanded coverage and particularly the use of references from international journals to support the statements made. In this particular proposal, there were a total of about 160 references of which only about 8 were to papers by the applicants of the proposal.

Introduction

Toxic blooms of a variety of organisms (harmful algal blooms (HABs)) have been documented throughout the world's coastal oceans, ultimately impacting shellfish, finfish, marine mammals and birds over large areas (Anderson *et al.*, 2002; Chen *et al.*, 2003; Dortch *et al.*, 1997; Field *et al.*, 2006; Fisher *et al.*, 2003; Janowitz and Kamykowski, 2006). Several such toxic species belong to the genus *Pseudo-nitzschia*, a group of marine diatoms that produce the neurotoxin domoic acid (DA) and are commonly found on the West Coast of North America, the Gulf of Mexico and in European waters (Anderson *et al.*, 2002; Dortch *et al.*, 1997; Lundholm *et al.*, 1994; Martin *et al.*, 1990; Trainer *et al.*, 2000). Not all *Pseudo-nitzschia* spp. produce DA, but over nine DA producing species have been identified. Coastal eutrophication is often cited as driving the observed increase in HAB frequency worldwide (cf. reviews by (Anderson *et al.*, 2002; Glibert *et al.*, 2005; Hallegraeff, 1993), but this may also be due to enhanced awareness and monitoring. In the case of *Pseudo-nitzschia* for example, a number of the species now recognized to be toxic were previously identified as "*Nitzschia seriata* (Cleve) Peragallo", a composite taxon that may contain both toxic and non-toxic species (e.g., Bates, 2000).

References provide support

Expanded coverage

The coast of California has been widely impacted by *Pseudo-nitzschia* blooms and the ensuing DA poisoning implicated in a number of marine mammal and seabird deaths (Bates, 2000; Bates *et al.*, 1998; Gulland *et al.*, 2002; Scholin *et al.*, 2000; Trainer *et al.*, 2000). Although the nearshore has been relatively well studied, the factors controlling toxic *Pseudo-nitzschia* blooms remain enigmatic. Several studies off central and southern California suggest that toxic *Pseudonitzschia* blooms are associated with the upwelling of colder high-nutrient waters (Anderson *et al.*,; Kudela *et al.*, 2004), whereas others report blooms after periods of high-nutrient river runoff (Anderson *et al.*,; Bates *et al.*, 1998; Dortch *et al.*, 1997; Fisher *et al.*, 2003; Pan, 2001; Trainer *et al.*, 2000; Van Dolah

References provide support

et *al.*, 2003; Wells *et al.*, 2005), or due to resuspension of seed-
ing populations into the euphotic zone during upwelling and
storms (Garrison, 1981; Trainer *et al.*, 2000). These studies are
further complicated by the fact that peaks in *Pseudo-nitzschia*
cell abundance and DA concentrations may be loosely cou-
pled in time. For example, increases in toxin production may
be a function of silicic acid or phosphate stress (Bates *et al.*,
1991; Fehling *et al.*, 2004; Pan *et al.*, 1996a; 1996b), as well as
iron and copper limitation that may occur prior to and after
Pseudo-nitzschia abundance maxima (Maldonado *et al.*, 2002;
Rue and Bruland, 2001; Wells *et al.*, 2005). The above hypotheses remain difficult to test due to a paucity of time-series
data.

The research idea and its context: Literature review

This section will give the author the opportunity to explain the research
question that is to be addressed in the proposal. As discussed in Chapter
Two, your research idea should be crystal clear and expressed in as simple
terms as possible, yet not oversimplified. You should be able to explain the
problem you are addressing in just a few minutes to someone who is not
an expert, as the reviewers may not all be specialists in the specific area
being described. Here is where you might apply the so called 'elevator
test' for a business plan — *viz.* you should be able to explain your problem
to someone in an elevator as you move between floors. This is normally
thought to take two minutes or less. It is worth spending some time on this
to get the level and the clarity right. **This initial part of your proposal
will create a first impression with reviewers** and will be your opportunity to engage their attention from the very beginning.

This section also gives you the opportunity to explain the background
of the problem using the published literature (see also Section 2.3),
including your own prior work if it is applicable. **Dealing with the published literature is an important step in preparing a proposal**. Your
review of the literature should demonstrate your awareness of the current
state of the field; should prove that there is a research gap that you will

address with your project; and should justify your proposed approach and your methodology. Provide references to justify claims that you make in the proposal, especially if these ideas call into question traditional approaches or orthodoxies. The more widely respected the authors and the journals are that you reference, the better.

> *There are only two publications cited, both about seven years old. While the applicants state that they have published related work to that described in the proposal, they do not provide detailed references to this work. Given the large amount of literature in the area of quantum chemistry and radiation damage in crystals, **the lack of a more extensive bibliography and references is a severe weakness of the proposal.***
>
> **— Quote from a reviewer**

One common, but fatal, mistake in any proposal is to state that **you will carry out a literature survey as the first step in your work plan after the proposal is funded**. A literature survey must be done **before** the proposal is submitted. Promising to do so later will be taken to imply that you are not really familiar with the current state of the field. Such a statement also undermines the confidence of the reviewers that there is definitely a research gap and that your research is truly novel.

By delaying the literature review you are also raising the possibility that **you may discover something that makes your proposal invalid, irrelevant or redundant if the work has already been done by someone else.** For all these reasons, reviewers will NOT want to see such a survey as the first action step of a proposal.

> *My problem with the time schedule as presented is that there is far too much time (2 × 6 months) spent at the beginning of the project in carrying out literature surveys. To my mind this should always be done as a prelude to submitting a proposal of this kind.*
>
> **— Quote from a reviewer**

In cases where the first language of the PI is not English, even if the proposal is written in English, there may be a temptation to refer mainly to local journals written in the PI's native tongue. Most international journals are now published in English, and most reviewers will not be familiar with local journals and so references to this literature will have less credibility with them.

Some of your reviewers may be leading researchers in your field. If they have written a relevant seminal paper and you have not cited it, this may annoy them. Furthermore, the reviewers will want to have confidence that you are familiar with the leading international journals and latest findings — once again, that you are aware of the 'state of the art'.

Finally, **you should NOT reference only, or even mainly, your own work** but instead indicate a broad familiarity with the field. This is the place where you can show for the first time that you understand the field very well, can clearly identify a research gap and show that your project will answer the question.

Methodology: Description of the research methods to be used

This section is in many ways the heart of any proposal and in our experience is **the piece of the proposal most often written badly and makes reviewers the most frustrated**. It is important in this section to describe in detail exactly **HOW** you are going to answer the research question you pose in the proposal. There must be sufficient information provided to allow the reviewer to judge whether your proposal is likely to be successful or not. Even if your technique is rather standard in the field, you should briefly outline your approach.

> **This is an example of part of the methodology for a successful BioMedical proposal to the US NIH. Note the level of detail provided. Note also the large number of references [numbers in square brackets] to other supporting publications.**
>
> **Measurement of Reactive Oxygen Species (ROS) levels and oxidative stress.** To substantiate and expand on results
>
> *(Continued)*

(Continued)

in other laboratories [41, 42, 44, 45], it is important to establish base-line data on ROS and oxidative stress levels in several cell lines. To do so, we will utilize several methodologies to measure the levels of superoxide and H_2O_2 in cell extracts. Superoxide will be assayed by nitroblue tetrazolium reduction [116], and will be expressed as pmol/min/μg extract protein. H_2O_2 will be measured by a spectrophotometric assay of the oxidation of 2',7'-dichlorodihydrofluorescein diacetate [117], and will be expressed in arbitrary units relative to the amount of extract protein. Since the specificity of this reagent in detecting H_2O_2 has been questioned [118], we will also utilize dihydrorhodamine 123, an analytical reagent that has been shown to be a more sensitive probe for the instantaneous detection of intracellular H_2O_2 by flourescence spectrophotometry [119]. Another alternative is peroxyflour-1 (PF-1), whose alkylboronate groups are selectively transformed by H_2O_2 to phenols, resulting in fluorescent products that can be measured by colorimetric or fluorometric methods [120]. Finally, a genetic sensor, the expression plasmid HyPer, will be utilized. HyPer encodes a circularly permuted yellow fluorescent protein (cpYFP) inserted into the H_2O_2-sensing protein OxyR; the plasmid is transfected into cells, and H_2O_2 production is monitored at the single-cell level by fluorescence microscopy [121].

Note the use of different techniques to establish results

Dealing with lack of consensus

For a project that is truly at the forefront of its field (or at the frontier of an entirely new field), you will be attempting something that has not been tried before. You will need to explain why this new method is likely to provide the answers you seek. If space allows, it is worthwhile comparing your methodology with alternative approaches and explaining why your choice is the most appropriate and effective for answering the research question and reaching the goals of the project.

One of our interviewees, whose research area is Public Health and Epidemiology, claimed that in his field, the methodology section of a proposal would typically take up about one half of the total proposal. Many of our interviewees, even those in very different research fields agreed generally with this opinion.

Particularly in the biological and medical sciences, it is becoming increasingly **important to have some preliminary data** that can help justify the proposed new work, and can demonstrate the feasibility of your approach.

You will need to **anticipate possible problems and explain how you plan to deal with these difficulties**. You may think that discussing possible problems weakens your proposal. However, if the reviewer sees difficulties that you have not mentioned, this could cause your proposal to be marked down. Better that you point out the potential problems and possible solutions as required. Having contingency plans will help convince reviewers that you have carefully thought through your plan of action.

> *There is not nearly enough detail provided in the research methodology to allow an accurate evaluation of the methodology nor its likelihood of success. For example, the discussion is all in terms of a general rare earth element 'R'. The number of different rare earth elements used is never stated nor which specific elements will be used. Another weakness is that there appears to be little fall back plans in the description of the methodology. What happens, for example, if the samples do NOT display the required superconducting properties? At a minimum, the proposers need to explain why they believe that there can be little doubt that the required superconducting properties will be achieved.*
>
> **— Example of a critical review because of the lack of sufficient detail in the methodology**

If your project requires special equipment, you should explain clearly where this equipment will come from. Will it be purchased as part of this project or does it already exist in your laboratory? If you are requesting funds to purchase special equipment, sometimes funding agencies will ask for one (or more) cost quotes to accompany the budget (see Chapter Nine). Alternatively, is a piece of equipment to be shared between many researchers, as would usually be the case where the equipment is very expensive? The arrangements for getting time approved to use the equipment, and whether time has already been approved, should be clearly spelled out in your proposal.

In some cases, your collaborators will be providing specialized equipment and again you should make it clear that they have agreed to your having sufficient access to it. If you have used this same equipment in the past. or if a collaboration using the equipment has occurred previously, this should be stated, since this will help to convince the reviewer that your project will be successful. In some cases, a letter from collaborators is required stating their willingness to participate. The letter should give clear details about what access the collaborators are providing and under what circumstances (see also Section 6.3). If such a letter is a requirement, this will be stated in the CFP. If it is not, you should check with the agency before adding it to your proposal.

You will also need to describe the timing of your project, since it is important to convince the reviewer that the project can be carried out in the time frame (and within the budget) available. A project schedule, or **Gantt chart**, is sometimes a useful tool in describing the time line for a series of interlocking tasks. It can illustrate how sub-projects feed into each other and how the outputs of one sub-project are inputs for another and so on. Even in simpler cases, some indication of the time frame for the different sub-tasks of the project needs to be given. Your goal, as always, is to provide enough detail to convince the reviewer that you have a good grasp of exactly what needs to be done and that you have set yourself a realistic timescale. Younger researchers in particular can be overambitious in terms of what is achievable. Unrealistic timescales can be very off-putting for reviewers.

The following Table is an example of a Project Timeline or simple Gantt Chart for a proposal with three objectives. The Chart is for three years, Y1, Y2, Y3 and has four quarters for each year, Spring, Summer, Fall and Winter.

Start Date: May 1, 2011		Y1				Y2				Y3		
	S	S	F	W	S	S	F	W	S	S	F	W
1st series of experiments (Objective 1)												
Sample analyses of 1st series of experiments												
2nd series of experiments (Objective 2)												
Sample Analyses of 2nd series of experiments												
Educational Outreach												
Data Analyses Synthesis (Objective 3)												
PI Meetings Conference presentations												
Publication												

Another common mistake is to combine a group of unconnected sub-projects into a single proposal. If you have two small but unconnected projects do not try to combine them into a single proposal. This will simply weaken both projects. You would be better to submit two separate proposals. The work plan should be an opportunity to illustrate how all the projects hold together and interrelate, making the whole greater than the sum of the parts. This is not possible with disconnected sub-projects.

Finally, if there are numerous investigators involved in the project, the responsibilities of the different people needs to be specified. If the group is divided into teams, these teams should be specified, along with each team's responsibilities in the project.

The difficult part of this piece of the proposal is providing enough detail about your methodology to convince the reviewer that the project is doable while not exceeding the space limits. **The methodology is an extremely important part of the proposal and you should not skimp on detail**. Reviewers will look very critically at the methodology section because this is primarily where they will determine whether you really can attain the goals that you describe.

> *The methodology is made clear by the use of appro-priate diagrams and figures. Plus there was a very good breakdown of specific tasks and a time line which seems realistic and appropriate. The level of detail is sufficient to be able to clearly follow what is planned and the plans seem realistic.*
>
> **— From a positive review by an experi-enced reviewer**

There is a myth in some scientific communities that by providing too much information you are giving away your important research ideas, which reviewers will then steal. In our experience this is very unlikely. We have not seen any examples of reviewers taking advantage of their role to steal ideas from proposals. As one of the experienced researchers whom we interviewed stated: "*If this is my one and only good idea, then just shoot me now*". Leaving the methodology unclear because you are afraid that your ideas or methods may be stolen is **not** a way to persuade a reviewer that your proposal will be successful.

Curriculum Vitae of the applicant and collaborators

Apart from a detailed description of how the project will be carried out, another excellent indicator of success is the past history of the applicant, particularly if the PI has worked previously in the area covered by the proposal. **This success can best be judged by the academic history of the PI, including publications in refereed journals, the outcomes from previous grants, a track record of working with collaborators, and experience with industry**, especially if this work is related to the topic of the current proposal. It is absolutely essential to include the CV of at least the applicant in the proposal.

There is a question as to whether the CVs of collaborators should also be included. This depends on how critical the collaborator is to the success of the proposal. This can be especially important in the case of interdisciplinary proposals (see Section 6.1.1), where it will be important to highlight key expertise that your partners are contributing in fields complementary to yours. If there are co-PIs, the same rule of thumb applies. The CFP should make clear whether collaborator and co-PI CVs are required, are optional, or should not be included. If there is uncertainty then you should check with the agency as to what is required and what is allowed.

It is also important to use the formatting suggested in the CFP in presenting your CV. Normally this will include your academic and work history, any prizes and awards you have received and a list of publications. Increasingly, in the interest of space, the reference list is restricted to a specific number of publications that are most relevant to the current project. In addition, there is often a space limitation on the total number of pages allowed for the CV.

While not usually quoted in a CV, some reviewers will look beyond the number of publications to the quality of the journal in which the publication appeared. One measure of quality is the 'impact factor' of the journal, reflecting the average number of citations that papers in the journal receive. In general, review journals have high impact factors. Biological and medical journals also tend to have high impact factors, reflecting the large number of scientists publishing in these fields. **Publishing in an internationally respected peer reviewed journal is very important in establishing credibility and productivity**.

Depending on the nature of your proposal and the case you have made for the relevance and impact of your work, you may want to highlight patents, activities with start-up companies, consulting, or industry partnerships in your CV. Such experiences are likely to give reviewers additional confidence in your ability to work with industry and your understanding of commercialization issues.

All CVs will not look alike, since more experienced researchers will be expected to have a longer, more extensive list of publications than a young researcher just starting out in his or her career. Reviewers are quite aware of such situations and can readily evaluate productivity appropriate to the stage in the PI's career. Some agencies elect to hold special competitions restricted to young researchers (see Section 11.4). However, our experience suggests that in most cases young researchers compete well with more experienced researchers provided that the reviewers take account of the appropriate background and experience of each applicant.

The evaluation of a CV can be more problematic with multidisciplinary proposals where reviewers from different fields often have different expectations regarding publishing rates and other productivity norms. For example, many more peer-reviewed papers per year are normally expected to be produced by biologists or chemists than by mathematicians. However, qualified staff of the funding agency will be aware of these issues and will take them into account. Panels with reviewers representing different fields can also help clarify the different expectations in different fields.

Another issue that may arise — especially in a country where the population of researchers in a particular field is small — is that identifying the applicants by including a CV with the proposal might introduce bias into the competition. Most applicants are very sensitive to the potential for bias and will often believe that it is the reason why their proposal was not successful. The solution to this problem is not to exclude the CV, which is so critical to the evaluation of the proposals, but instead **to use reviewers from outside the country who would not be biased for or against particular applicants** except in very unusual circumstances. Of course, applicants generally have no control over the selection of reviewers (see however Section 3.2 for the case where applicants are asked to submit names of potential reviewers) but in the case of proposals submitted in fields with small numbers of local reviewers available, you should be aware that the agency will be likely to select international reviewers, and you will need to be particularly careful in explaining any unique aspects of the local situation.

Budget

The budget is also a critical piece of the proposal. In most cases, it is more important to the staff of the agency, who are aware of the financial constraints within the agency and who must make priority decisions about overall budgets. Even so, reviewers can provide insight into the cost of equipment, supplies, and travel as well as guidance on the number of people needed to carry out specific tasks. We will deal with the budget in more detail in Chapter Nine. Suffice to say here that the format described in the CFP should be carefully followed in the presentation of the proposed budget. The staff of the grants administration office of your institution can also be very helpful in giving advice since they deal with grants and budgets on a regular basis.

4.3 Use Figures and Diagrams for Clarity

Even though you have only a limited amount of space to present your proposal, you will often find that using figures, charts and even photographs will make your proposal clearer and more easily readable by the reviewers. The old adage that a picture is worth a thousand words is usually true. A figure can be especially useful when you are trying to explain a complicated experimental setup. Just be sure to label the figure clearly with labels large enough to be readable.

The example shown below is taken from a geology proposal, 'Space-time variability in seismic and aseismic strain release at a trench — aseismic ridge intersection: CASA GPS Geodesy in the North Andes' (thanks to Professor James Kellogg) discussing measurements of plate tectonics near the Ecuador–Columbia trench. The figure shows the complicated movements in the region in a way that would be almost impossible to describe in any other way.

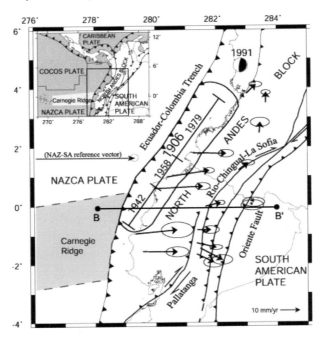

Comparisons between experimental measurements and theoretical predictions are also often best presented in a diagram. It can be used, for example, to illustrate how the measurements you expect to make can distinguish between the predictions of different theories.

Complicated chemical formulae can also be presented more clearly using a diagram, such as the one below taken from an actual proposal showing the structure of certain amidine derivatives (thanks to Professor Daniel Reger).

A

B

There are many similar examples in many different fields of science where figures clarify or illustrate a situation. You simply need to be sure that you explain the figure in enough detail that it adds to the understanding of the point you are trying to make.

4.4 Include Only Required Material

As outlined in the previous section, the CFP will require certain information in the proposal. It may be tempting to provide additional material that you believe will be supportive of your proposal, but this should only be done with extreme caution. In the first place, there is the danger that, if too much additional material is included, the proposal word limit is exceeded, making the submission ineligible. Even if the material does not result in the proposal being rejected, reviewers do not normally appreciate having to deal with more material than required by the CFP.

One particular example of extraneous material that accompanies proposals is letters of support. While it may be useful to have a letter from a collaborator who will indicate her or his willingness to participate and will describe the role that she or he will play in the project, there is no reason to include letters from other people who are simply advocating support of the proposal. Reviewers view the strongest letters of support as those that specifically contribute tangible assets to a proposal, e.g., money, people, equipment. Most, if not all, reviewers will look on unsolicited endorsements with a jaundiced eye and these will not help your proposal.

> *I don't see the point of extra 'support' letters that all look as if they were written by the applicant and that provide no specific resources either in equipment or personnel to the project.*
>
> **— Quote from a reviewer**

However, there are situations where a letter or letters of support may be required. Competitions which seek university–industry cooperation often require letters from the industry partners. It may also be critical to include a letter of support if, for example, you have made any claims about an important contribution from your partners which is a key selling point of your proposal, such as any significant financial support from an industry partner or access to critical equipment from an academic partner.

4.5 Institutional Approval

In most research institutions, some form of administrative approval will be required before a proposal is submitted. This may involve an actual review of the proposal itself either internally or externally. At a minimum, there will almost certainly be a sign off required by the research office (Vice President for Research or her or his representative) to check for either monetary or space commitments in the budget and that the proper indirect costs have been included in the budget (see Chapter Nine). Such signature requirements are another argument for an early start to preparing your proposal and for not leaving the submission till the last minute.

There are **special approvals required when either human or animal subjects are involved in the proposed research**. The requirements associated with these approvals will vary from country to country but the general principles are similar. In the case of human subjects, the goal is to protect them from harm, and to ensure that they give informed consent to their participation. For the use of animals, particularly vertebrates, the aim is to ensure the humane care and treatment of animals used in research. These requirements are normally overseen by review boards in each research institution. In the USA, for example, any research involving human subjects must be approved by an Institutional Review Board (IRB), which will typically review a proposal before it is submitted to a funding agency. Similarly, for any research involving animals, there is the Institutional Animal Care and Use Committee (IACUC) whose approval is required. Other countries have similar organizations with similar requirements. Therefore, for any research involving humans or animals, approval is required before your research can begin or in most cases before the proposal can be submitted. Once again this is an argument for making sure you leave plenty of time to obtain the required approvals before the deadline for submission of the proposal.

Dr Frankenstein was disappointed with the letter from the Ethics Committee.

4.6 Write for the Reviewers

As we explained in Chapter Three, reviewers are usually active researchers in the general field of the proposal. In cases of large grants with comparatively few proposals submitted, it is possible for the staff of the agency to seek reviewers who are quite specialized in the area of your particular proposal. This is quite an expensive and time consuming process. In other cases, where the number of proposals submitted to the competition is large, the reviewers of your proposal will not be experts in your particular specialized area. This is likely to be even truer for multidisciplinary proposals that do not fit neatly into one particular area of science or engineering, but cross between different fields.

Try to explain your project clearly, to minimize the use of jargon and not use acronyms known only to people in the field.

Not everyone has the ability to set their idea on paper clearly and succinctly. What you are trying to say may seem very clear in your own mind but be less clear when written down. Here are two pieces of advice that may be useful. One is simply to **read what you have written out loud**. Sometimes hearing an idea will help you to clarify your thoughts. The second is to **have someone else read your proposal**, especially if they have experience in reviewing proposals. Choose someone who is a critical reader and will tell you the truth as they see it. You do not need to follow their advice, but it is very useful to get a critical reviewer's input on what you have written.

In cases where the proposal must be submitted in English (especially if English is not your native language) you must be very careful to check your grammar either with a native English speaker or seek help from your home institution in obtaining help to do so. This may not seem very important since the idea behind the proposal is the most important thing. But, if your command of the language and your mastery of its grammar are poor, the reviewer will have difficulty understanding what it is you are trying to explain.

> There is the classic example of the book "*Eats, Shoots and Leaves*" by Lynne Truss, where the extra comma after Eats makes the phrase both ambiguous and very amusing.

This leads to another basic principle: **Do not annoy the reviewers!**

There are many ways you can do annoy them — from poor grammar and the inclusion of extraneous material, to poor formatting and a font that is difficult to read. If a reviewer has to use a magnifying glass to read your proposal, be assured that he or she will not look kindly upon it. Make sure that the logic and coherence of your argument is sound, and that the different pieces connect together seamlessly. You should try very hard to make your proposal as clear and as readable as possible.

Remember, the reviewers will normally read a number of proposals and they will be reading them critically and seeing them as part of a competition. A spelling or grammatical error on the first page will often be interpreted as carelessness on the part of the applicant. Check your proposal carefully and make sure that any simple mistakes are caught and corrected. This may seem like very obvious advice, but is neglected by a surprisingly large percentage of applicants.

4.7 Summary

This chapter discusses the mechanics of actually writing a research proposal and provides advice on the various sections of a typical proposal. Remember that preparing a proposal takes time, possibly as long as a year, so you should make sure to leave yourself plenty of time to do a careful job.

One of the most important pieces of advice is to read the CFP very carefully and make sure that you address all the criteria listed there. Make certain that you follow the suggested fonts and formatting and keep within any page limitations for the total proposal and any subsections, especially the CV and Budget sections. Most importantly, make sure that you submit the proposal by the deadline and remember to allow time for both the revision and any administrative delays that may arise in your own institution.

In the **abstract**, briefly present what the proposal is about, why it is important and how you plan to carry out the project. This should include any special equipment required, and any collaborations that will be involved. The **introduction** will expand on the abstract with more detail particularly about the novel aspects of the project. The **literature review** should place your proposal in context, identify a research gap and explain how you will answer the research question. In the literature review, make sure to refer to international peer reviewed journals and not only local journals. This will determine for reviewers how familiar and current you are with the field of the proposal.

Don't refer only to your own work in the literature review but make sure to include all the relevant papers. It is a mistake to say in your proposal that you will begin with a review of the literature. This should always be done before submitting the proposal.

In the **methodology** section, you will need to explain in detail how you will carry out the project. You should discuss possible problems that could arise, and how you will deal with them. Providing a time frame for the different pieces of the project will add credibility provided it is realistic. You should describe the role of any collaborators and explain the arrangements for access to any pieces of specialized equipment needed for the project. In the CV section, include not only refereed peer reviewed publications and previous grant funding as required but also experiences with prior collaborations or industrial connections as appropriate. The budget should include a justification particularly for any major expenses. Only include relevant material and do not clog up your proposal with extra material that is not required such as generic letters of support. At best, reviewers will simply pass over such material.

Remember that you are writing for a group of reviewers and you must take care not to do anything that will annoy them. This means writing as clearly as you can with a minimum of jargon or acronyms that are only common in your specific scientific or engineering area. Be particularly careful to use good grammar and correct spelling. Get help if your English is not colloquial. Such mistakes are common, but quite easily avoided.

DO start in plenty of time to prepare your proposal.

DO read the CFP very carefully and follow all the recommendations regarding formatting and fonts.

DO cover all the criteria listed in the CFP.

DO submit the proposal by the deadline.

DO include WHAT the project is about, WHY it is important and HOW you will carry it out in the abstract.

DO explain how it fits into current work in the introduction.

DO include all relevant references in peer reviewed international journals.

DO explain in detail how you will carry out the project including how you will deal with possible problems that might arise.

DO NOT exceed the page limits either on individual sections or on the proposal as a whole.

DO NOT include only local references or only your own work in the literature review.

DO NOT include extraneous material that is not required in the CFP.

DO NOT annoy the reviewers by using poor grammar or spelling or by using jargon or undefined acronyms.

Chapter Five

Re-Drafting the Proposal

Good writing is bad writing that was rewritten.

— Marc Raibert

Great work! You now have a first draft of your proposal, and are congratulating yourself that it is completed before the deadline. Unfortunately, this is only the first, albeit a very important, step.

As we mentioned in the previous chapter, it is vital to leave plenty of time before the deadline to begin preparing the proposal. One reason for requiring plenty of time is that, after having completed a draft of the proposal, the extremely important process of **re-drafting** begins. This is necessary in order to prepare a final version that can be submitted with a reasonable chance of success. This process takes time. All the more so when research collaborators or other partners may be involved in shaping the proposal, or need to be consulted on particular aspects of the project.

> *You cannot wake up one morning and write a proposal; it needs substantial preparation.*
>
> *The later you start, the less likely your proposal is to succeed!*
>
> **— Quotes from two experienced reviewers**

In our view, and it is a view that is shared by almost all the experienced grant-writers whom we interviewed, **having critical readers review the first draft is essential in preparing a final proposal**. In this chapter, we will discuss various approaches to re-drafting, including who should read the first draft, what they should be looking for, and how you should respond to their comments.

5.1 The Author Should be the First Reviewer

This may seem like an obvious point. You are the person who is most familiar with the content of the proposal, with what you intend to do and why this is important. Therefore, it makes sense that **you should be the first critical reader** of the draft of the proposal. You will need to set the original draft aside for a day or two so that you can approach it with a fresh perspective.

Take time to read the Call for Proposals (CFP) carefully one more time and identify the goals and criteria for this competition. Try to look at the draft **not as the author but as a critical reviewer**. In some cases, the funding agency may provide a reviewer form, including a scoring scheme. Or a colleague who has previously reviewed for the agency may be able to supply a reviewer form. It is well worth scoring your own proposal according to the guidelines supplied by the agency.

> *I have the impression that the proposal was put together in a hurry. For example, paragraph four of Section Three. 'Methods of Research' is left completely BLANK implying a **lack of any, never mind careful**, re-reading of the proposal.*
>
> **— Quote from an experienced reviewer.**

You will need to bring a degree of objectivity to the proposal that may be difficult. Ask yourself questions like:

- Does the proposal, as written, meet the formatting and space constraints of the CFP?
- Is it clear **WHAT** research question this proposal is asking?
- Does the proposal make it clear **WHY** this question is important?
- Is the proposal too broad, too unfocussed or too vague?
- Is the novelty/uniqueness of the proposal spelled out clearly?
- Does the proposal address all the criteria of the competition as laid out in the CFP?
- Is it clear **HOW** the proposal will answer the research question posed?
- Has the proposal dealt with any problems that might arise in carrying out the project?
- Does the reference list help amplify and clarify any points that might be uncertain in the proposal?
- Does the budget ask for what is needed to carry out the project?
- Is the format of the Principal Investigator's CV consistent with the requirements of the CFP?
- Are the CV's of any collaborators provided, if required?

You will almost certainly find that, if you are objective about re-reading your first draft, there are improvements that can be made. Once you have made the changes that you believe are necessary, you are ready to move on to the next step.

5.2 Review by Colleagues

Choosing critical reviewers who can help improve your proposal is extremely important. One source is colleagues who work reasonably closely with you, even if not in the same research group. These colleagues can make valuable reviewers since they will be familiar with the subject matter of your proposal and the relevant research domains and so can give critical and informed guidance.

Some colleagues may even have acted as reviewers for the relevant program/competition or have been part of review panels. These colleagues can be especially helpful as readers of the draft of your proposal.

> *If you are the only person who has read the proposal before it is submitted, you have little or no chance of success.*
>
> **— Quote from a reviewer**

In some cases, research groups or academic departments will have a formal review process for all proposals. Sometimes this happens early, at the idea stage, when people are encouraged to present their ideas for a proposal to the group for input. Later, when the proposal is drafted, members of the group can be asked to read over the draft and provide a critique. Alternatively, there may be an informal arrangement for new members where senior members of the department or group agree to read over and comment on proposal drafts. You should certainly take advantage of this option if it is available.

While using work colleagues as reviewers is usually the simplest and easiest option, it does have some disadvantages. Your colleagues may be too close to the work and may make the same assumptions that you do in presenting your proposal. These assumptions may not be shared by the reviewers chosen by the funding agency, who may not be precisely in the same field. What may be clear to colleagues in a very closely related field may not be so clear to a reviewer who is not a specialist in this particular science or engineering area.

> *In some research groups, the head of the group will not sign off on a grant submission unless some other members of the faculty have read it and given advice to the applicant.*
>
> **— Comment from one of our interviewees**

In addition, colleagues and friends may find it difficult to be sufficiently critical of your work and may 'soft-pedal' any criticism. They may not want to hurt your feelings since they have to work with you regularly, and being critical of your proposal may make this awkward. You need to develop a thick skin that will enable you to encourage and accept honest, well-intentioned, criticism and even appreciate the value of it.

> *Don't fall in love with your idea; be prepared to modify it if need be.*
>
> **— Advice from an experienced proposal writer**

You also need to be aware that it is important to listen carefully to the criticism of colleagues, and not to be defensive about whatever improvements they suggest. Discuss points of disagreement with them. Remember you do not have to change the proposal in response to criticism but you should take the comments seriously even if you decide not to make all the suggested changes.

If your colleagues identify what they see as weaknesses, then the actual reviewers may well do so also. Therefore, even if you disagree with your colleagues, the process of independent review can help identify areas where you might usefully strengthen your case — providing pre-emptive counterarguments on issues which might otherwise be disputed.

5.3 Using Readers Who are not Colleagues

There are advantages in using people who are not close colleagues as reviewers. First, they are likely to be more objective since they do not deal with you on a regular basis, and they will treat what you have written with a more objective and critical eye. Also, they are likely to be less familiar with the details of the work and so will be in a better position to determine whether your description, both of the problem and how you will tackle it, is presented clearly. In other words, these reviewers can help you identify what might be less clear to the more generalist reviewer.

> *Vet the proposal with as many colleagues as possible before submission. One of the problems people have is embarrassment about their writing and they are reluctant to share their work with other people.*
>
> **— Quote from a reviewer**

One problem you may find with using multiple reviewers is that you get contradictory advice from different reviewers. Our advice is to wait to get all the feedback from the various readers and then take time to consider all the inputs before re-drafting. In the end, you need to make the decisions since, after all, it is your proposal.

The ideal critical reader of your first draft is someone who is most like the reviewers who will be reading your final, submitted proposal. Therefore, you should think about who these reviewers are likely to be, realizing that they may not be precisely in your field. Perhaps in your department there are scientists in neighboring fields who might be willing to read your proposal and give you some opinions.

> *Leave your ego out in the hallway and listen to constructive criticism.*
>
> **— Comment from a laboratory director to his staff**

In some institutions, the research office (Vice-President for Research or some similar office) will offer a service of finding, and sometimes even paying, external reviewers to read over drafts of proposals and provide comments. If this service exists in your institution, be sure to make use of it. Offering even a modest stipend for a reviewer will often mean that the person will take the task more seriously and so it is worth a small investment. Realize, however, that any review, especially an external review, will take some time. So, again, this is an argument for starting early in preparing your first draft. If your institution does not have such a service, you could still ask the research office for help in finding reviewers in particular cases.

While program officers from funding agencies are not allowed to review proposals before they are submitted, it is possible to get general advice about research ideas and the current goals of the agency that may affect your writing. However, retired program officers are allowed to review proposals since they are no longer associated with the funding agency. Because of their extensive experience, they make excellent reviewers of draft proposals.

5.4 English Language Reviewers

There is a special case for re-drafting in countries where the language is not English and yet the proposals are prepared and reviewed in English. In such cases, review of the first draft of the proposal is even more critical. We have read a number of proposals where there appears to be the kernel of a good idea but the English is so poor that it is impossible to be sure and the proposal therefore does not receive a good score. The best choice of a critical reviewer of the draft in this case is a person whose native language is English or at least someone who has used English professionally for some considerable period of time.

> *The level of English in this proposal is so poor that it is very difficult to understand just what the authors are proposing and how they mean to carry it out. It is necessary for me to repeatedly guess the meaning of a sentence.*
>
> **— Quote from a reviewer**

The research office of your institution may be helpful in finding English language reviewers. While it is preferable to find reviewers who work in areas close to your research, it is possible to make use of readers who do not know the domain well, but can still read over the English to make sure it is grammatical, clear, and free of spelling errors. Language departments in your institution, if they exist, may be another source of possible English language reviewers.

> *The English is a major problem in this proposal. For example, what does the following sentence mean? 'The roughness is necessary for modeling spherical ledges, where after the loading of the contacted body will approach for some size, also other segments of a rough surface will come into contact'.*
>
> **— Quote from a reviewer**

Poor English or English that is unclear is another way in which **you can annoy the reviewers of your proposal** and, as we have said before, **this is certainly something you do not want to do**.

5.5 Summary

Now that you have written a first draft of your proposal, the hard work is just beginning. You will need to re-read your proposal carefully making sure that it deals with all the criteria listed in the CFP. You also need to check that it conforms to all the formatting and page requirements of the CFP. All this takes time which means that **you need to start preparing your proposal well before the deadline**.

Once you are satisfied with the proposal, you then need to have it reviewed by other people. Your close scientific colleagues, especially ones who are **known to be critical or are themselves successful proposal writers, make excellent readers**. If you have access to readers outside your local environment that is even better since, in general, they can be more objective about your writing. Ideally, the best people to read a draft of your proposal are people who match closely the group of reviewers who will finally review your proposal for the funding agency.

Most important, if English is not your mother tongue but the proposal must be submitted in English, make certain that **someone who has an excellent familiarity with English reads the proposal to correct any grammatical or spelling errors** which might make the proposal difficult to read.

DO leave plenty of time for writing your proposal before the deadline especially to allow time for re-drafting.

DO have at least two or three colleagues read over your draft and make suggestions for any improvements.

DO have people from outside your immediate colleagues read your draft for clarity if this is possible.

DO take advantage of any program organized by your department or institution to review your proposal externally before submission.

DO make sure that someone with an excellent command of English checks your proposal for grammar and clarity.

DO take any comments or suggestions seriously although you do not have to incorporate all of them.

DO NOT be the only person to read your proposal before submission.

DO NOT feel threatened by critical comments but value them since they will likely improve your proposal.

Chapter Six

Partnerships

Science is a collaborative effort. The combined results of several people working together is often much more effective than could be that of an individual scientist working alone.

— John Bardeen, Nobel Laureate for Physics

6.1 Introduction

While it may not apply to all proposals, it is certainly worth considering the idea of linking up with other people to broaden the scope of your project. Such cooperation may add value and make your proposal more competitive.

There are many ways partnerships can enhance your planned research, whether by filling critical gaps in your expertise, accessing useful equipment or data, or gaining insights into the potential applications of your work. When designed carefully and articulated clearly such partnerships may significantly strengthen your research proposal in the eyes of reviewers and funders. In this chapter, we share lessons and practices for communicating the nature, value and organisation of your partnership.

In addition to the core mission of funding excellent research, many national research foundations often include partnership development as one of their goals and, consequently, may prioritise support for activities which do some or all of the following: increase interaction between academic researchers and industrial researchers; address multidisciplinary research challenges through partnerships; grow the level of collaboration between different national universities (and other public research institutions); and strengthen connections with the international research community.

Different research projects and funding programmes have different requirements for collaboration and partnership. In this chapter we pay particular attention to the opportunities (and challenges) of establishing partnerships with:

- Researchers from **other disciplines**.
- **Industrial** collaborators.
- **International** researchers.

Before you invite collaborators to join your grant proposal, however, it is worth pausing to consider if you really need them as partners or whether the insights, expertise or access to facilities can be obtained in other ways. For example, it may be sufficient to have experienced researchers on an 'advisory panel' for your project. Established researchers are often willing to assist colleagues in this way, especially if they are helping younger colleagues or those based in less well established research environments. Similarly, it is sometimes more efficient and straightforward to buy access to equipment or services at specialist facilities or research and development (R&D) institutes, rather than gaining access through collaboration.

When describing the contribution of your research partners in your grant application, you will need to be able to:

- Explain how the partnership will provide **added value** to your research plans.
- Demonstrate that you have the capacity to **manage** the partnerships.
- Convince your reviewers that the proposed partnerships are **real and sustainable**.

You will, of course, also need to be able to explain these factors to your potential partners when you are inviting them to join your proposal! This chapter addresses a range of themes and issues which are important in effectively communicating the value and feasibility of your partnerships:

- The added value contributions of your partners to the proposed project: *What do they bring?*
- The challenges of managing collaborative research: *How will you work together? How will you resolve disagreements?*

- The commitment and engagement of your collaborators: *Are these partnerships for real?*

6.1 Added Value

In this section we share lessons and practices on how to communicate different types of added value which a partnership might bring to your grant proposal. In particular, we explore the opportunities and challenges associated with communicating different types of contribution.

One of the main reasons researchers invite partners to be part of their research proposal is, of course, to strengthen their proposals by filling gaps in their **knowledge and experience**. This knowledge can come in a number of forms, for example:

- **Domain knowledge**: Additional scientific expertise, in particular from complementary research domains, required for multidisciplinary research.
- **Application knowledge**: User insights, which can enhance the potential for translation and impact.
- **Added experience (and credibility)**: Research insights and track record of successful research projects.

Partners can also provide more practical support by sharing their **technical infrastructure, resources and capabilities** in a variety of ways, for example by providing:

- **Access to know-how**: Experimental techniques, skills associated with particular methodologies, experience with types of equipment.
- **Access to facilities**: Equipment or major infrastructure (e.g., synchrotron facility).
- **Increased research capacity and efficiency** through pooling a critical mass of expertise, effort, and effective division of labour.

In the following sections we provide examples of what reviewers look for in research partnerships that are providing added value by (1) adding complementary knowledge and expertise; and (2) providing additional research capabilities.

Partners providing knowledge and experience

If your proposed project is in a research domain that is relatively new for you, or if you are using new methodologies or techniques, there can be particular value in collaborating with researchers who are experienced in the field. This helps give reviewers the confidence that you are likely to make an effective transition into this new area and that your project is likely to be successful.

> *The involvement of the impressive list of multidisciplinary collaborators is what sets this proposal apart.*
>
> **— Quote from a reviewer**

Working with your research partners to gather preliminary data in advance of submitting your grant proposal is often a good idea. Not only is preliminary data helpful in its own right, but the early collaboration indicates to the reviewers that you have an established and working relationship with your research partners.

> *Not only does the preliminary data look extremely promising, but the fact that both teams put such a significant combined effort into running the experiments suggests considerable joint commitment, as well as providing evidence of an effective working relationship.*
>
> **— Quote from a reviewer**

Partners contributing domain knowledge

Reviewers will quickly make a judgment, based on your research question and CV, as to whether you have the background and track record to deliver on your stated project goals. Sometimes you may attempt to take on a research challenge that is not well aligned with your CV, either because you are moving into a relatively new field for you or because it requires input from a number of complementary fields. Under these circumstances it can be worth finding research partners who have experience and a track record in the relevant areas. In addition to their expertise, they may also be able to provide data, materials and samples, etc.

> *The collaboration with a leading microfluidics expert was a smart move. This was a missing piece of the jigsaw. This gives the core group of device-physicists and clinician-researchers a real platform to develop novel biomedical diagnostic devices.*
>
> **— Quote from a reviewer**

Partners providing application knowledge

Some research partners — especially industrial collaborators or more applied academic researchers — can offer insights related to the next phase of the innovation journey from idea to application. Not only can this type of knowledge help you overcome challenges related to translating more fundamental laboratory research into increasingly applied systems and environments, but it can also increase the likelihood of impact in the real world.

> *The proposal seems a little naive regarding the difficulties of turning [recent applied science findings] into a viable technology. Nevertheless, the close involvement of [the industrial partners] should help ensure the research plan is informed by real-world commercialization and scale-up challenges.*
>
> **— Quote from a reviewer**

> *The huge volume of real-time data the industrial partner is able to provide should prove invaluable in informing and testing the network model.*
>
> **— Quote from a reviewer**

Industry partners can offer more than just user insights, real-world data or advanced technical capabilities. Industry partners can also help identify important **and intellectually interesting** industrial problems. These can be invaluable in helping you identify new research projects or enhance the translation of your research knowledge into industry. Some major R&D corporations or highly innovative smaller firms may even be right at the forefront of some applied research fields (or have the most advanced equipment or techniques) which can add huge value and competitive advantage to your research endeavours.

STORY: How the industry partners made the proposal stronger e.g., helped identify an important real world application of the research.

When an applied physicist approached a major information and communications technology (ICT) company to explore potential applications of his research to telecommunications devices, the company's research team identified an even more promising medical device sensor application. The company then introduced the researcher to their clinician collaborators and offered access to their medical device 'test bed' as a cost share contribution to the grant proposal.

The reviewers singled out the novelty of the application, its potential impact and the quality of the industrial engagement for particular praise when recommending that the proposal be funded.

Collaborating with industry partners can also provide valuable research experience to graduate students and post-doctoral researchers. In particular, early career researchers get exposure to the research culture and practices of the private sector, gain awareness of career opportunities in industry and experience the satisfaction of working on real-world problems.

Partners providing additional experience (and credibility)

If you are an early career investigator, are entering a new field, or are coming from a less mature research environment then collaborating with experienced researchers (who may be well known to the reviewers), can significantly increase the confidence of reviewers in the probability of your project delivering on its goals.

> *The level of support for this project by [such a high profile collaborator] suggests they believe the applicant has what it takes to deliver.*
>
> **— Quote from a reviewer**

Things that may help convince reviewers

- **DO** be clear and specific about your partner's contribution to your research project.
- **DO** make sure your collaborator's resumé highlights the areas of knowledge and capability relevant to your proposed joint activity.
- **DO** use the literature review section of your proposal to illustrate the relevance of your partner's research (perhaps even directly citing their work, if appropriate).

Partners providing new capabilities

If your proposed project is in a research domain that requires a complex or elaborate technical infrastructure, sophisticated skills and techniques, or a critical mass of researchers from diverse backgrounds there can be particular value in pooling skills, equipment and resources. This can give reviewers significant confidence that you (and your partners) can be successful.

Partners contributing know-how

Knowledge of the latest relevant literature and techniques is, of course, often not enough to address your research question. Sometimes the ability to deploy this knowledge is embedded in tacit knowledge of an experimental technique that is difficult to codify, hard to learn, and can often only really be shared by regular practical engagement and trust. This kind of know-how can be particularly important for successfully deploying emerging experimental techniques or particularly complex methodologies. Experienced reviewers will be aware if your proposed project is likely to face additional challenges of this type, and the implications for its success and competitiveness. Finding partners with a track record which suggests they have this kind of know-how within their team can be invaluable.

> *Collaborating with Professor X's group, and in particular Dr Y, gives this proposal a huge advantage. Dr Y's involvement in the original development of the probe itself means there are few people better placed to perform this complex experiment.*
>
> **— Quote from a reviewer**

Partners providing access to facilities

One particularly valuable contribution a research collaborator can make to your project is in providing access to specialist equipment and infrastructure (and the staff to help you use it).

> *Access to the collaborators' facilities — which may be the most sophisticated in the country (if not the world) gives the applicant a major competitive advantage.*
>
> **— Quote from a reviewer**

If your partner's equipment is a source of significant competitive advantage, make sure you let the reviewers know this. And give them enough information to understand why this will allow you to do things other researchers cannot.

Partners increasing your research capacity and efficiency

If you are tackling a research challenge that requires more expertise, effort, and/or critical mass than could be achieved by your team alone, partnerships can increase the effectiveness and efficiency of your endeavours through division of labour. This can be particularly important in fast moving and competitive research fields.

> *The proposed project is taking on an immensely complex, labor-intensive challenge in a highly competitive area. The effort he has gone to in bringing together a national 'consortium' with the critical mass to take on the challenge is extremely impressive.*
>
> *I reckon he deserves the opportunity.*
>
> **— Quote from a reviewer**

Things that may help convince reviewers

- **DO** give sufficient detail to reveal what makes your partner's know-how exceptional, e.g., experienced 'super technician'.
- **DO** give reviewers enough information to understand why any shared equipment will enhance your competitiveness.
- **DO** provide any details of how the partner's contribution will help achieve project goals more efficiently.

6.2 The Challenges of Partnership: Management and Planning

Research partners and collaborators can help you pursue more ambitious, complex and impactful research. With this potential, however, come additional challenges for planning and managing your research. In this section we share lessons and practices for how to give reviewers assurances that your collaborative research grant will be managed effectively. In particular, we explore ways of giving reviewers confidence that you have:

- **Anticipated the challenges associated with collaboration including** multidisciplinary, international and industrial partnerships.
- Put in place **engagement, governance and management plans** to deal with these challenges.

Anticipating practical challenges associated with collaboration

Working effectively with research partners can sometimes be challenging, especially when those collaborations are new. Experienced reviewers will anticipate these challenges, and may be less convinced by proposals where plans for collaboration appear naive.

> *The schedule of the collaborative work programme looks extremely ambitious for a team that have never worked together before... I am unconvinced they can pull this off.*
>
> **— Quote from a reviewer**

When writing a proposal it is a good idea to try and anticipate these challenges and, where possible, put plans in place to mitigate them. Potential challenges include:

- Your partners may have other projects and commitments that are priorities for them. If the reviewers believe your partners are likely to be too distracted by other priorities or too busy to commit the amount of time implied by the proposal, this will undermine their confidence in your planning and your project.
- Your partners may have existing plans to use their equipment. Reviewers may be suspicious of plans which imply access that appears too easy or even just too regular.
- It typically takes time for new collaborations to learn how to work together effectively. Extra effort will probably need to be invested in communication (and potentially resolving conflicts and tensions).
- Joint decisions on revising the research agenda will require consensus and may take longer than you anticipate.
- If you are an early career researcher or are coming from a less well established research environment, it can be important for your career to develop an independent research identity. In particular, your proposed project should not look like an extension of your research partner's agenda. The proposal should make clear that the project has been

conceived by you and will be driven by you if funded. This will be especially important if your collaborator is a former research supervisor.

- It may be especially difficult to shut down sub-projects which are not working if they involve your partners. There is a risk that you will get 'locked in' to the original proposed research agenda and being unable to revise the research plan to take advantage of new insights, priorities or opportunities. Project plans with clear governance structures, review milestones and transparent decision criteria can be helpful.

> *The work schedule implies the team will have access to Professor X's pilot scale bioreactor for much of year 2. This seems unlikely, given the grant commitments listed in Prof X's resume.*
>
> **— Quote from a reviewer**

Things that may convince the reviewers

- **DO** highlight any evidence of experience in participating in (or, even better, managing) collaborative research projects.
- **DO NOT** overstate the time commitment of your partners.
- **DO** put in place processes and governance mechanisms for revising the research agenda (see section on Management and Governance below).

In addition to some of the general challenges of managing an effective partnership outlined above, different types of partnership will have their special issues and difficulties. The following potential challenges should also be considered when preparing your proposal. Where possible, you should try and give the reviewers confidence that you can overcome them.

Challenges associated with multidisciplinary partnerships

Some of the most exciting and important research challenges lie at the interfaces between traditional academic disciplines. With these opportunities, however, come challenges. In particular, your research partners from other fields may use different language and terminology, may want to publish in different journals, draw upon different literatures, use different techniques and sometimes value different methodological approaches. When writing your grant proposal it can be extremely helpful to refer to journals, discuss methods, and use (and 'translate') terminology from all the fields relevant to your proposal.

"Maybe we should reconsider a research collaboration with the Engineering Department...."

Multidisciplinary Partnership Example: Nanotech and Biosensors

Partners with the right combination of equipment, data and techniques can be invaluable.

A proposal from researchers who had identified a novel application of quantum dot-based sensors for healthcare (related to monitoring of food products for toxic chemicals) impressed the reviewers. The reviewers were concerned about the complexity of some of the research challenges in areas they felt needed to be supported by high quality technicians and specialised laboratory set-ups.

The reviewers were, however, "*reassured by the very helpful and substantive letters of support from a leading US research institute with particular expertise in the type of multidisciplinary research in-keeping with the proposed problems*", which offered "*real resources, for example, access to key equipment, data bases, and expert technician time, etc...*"

There may also be some practical grant application challenges associated with multidisciplinary projects. For example, proposals are likely to be sent to reviewers from different disciplines and so you will need to write the application using language and terminology that will be understandable by all the reviewers.

> *The use of jargon is excessive. I simply cannot judge whether the outputs of the materials science sub-project will generate results relevant or useful for the sensor design work.*
>
> **— Quote from a reviewer**

Communicating to a multidisciplinary audience can be made all the more challenging if the funding agency imposes the same word limit as it does for a single discipline proposal. Within these constraints, you will need to convince reviewers that you know enough about other disciplines to have selected the right partners and will be able to communicate and collaborate effectively.

Things that may help convince reviewers:

- **DO** include a literature review which appropriately addresses all relevant fields.
- **DO** write the application using language and terminology that will be understandable by reviewers from different disciplines.
- Where possible, **DO** include plans for team members to spend time in your collaborators' laboratories.
- Where possible, **DO** hire (or indicate an intent to hire) team members with some experience in the research field of your collaborators.

Challenges associated with industrial collaboration

Collaborating with industrial partners can enhance the quality, relevance and impact of your research, especially in many technology- or management-related research domains.

There can, however, be significant challenges to developing an effective partnership with industry. For example, your industry partners will often be driven by different motivations, project timescales and constraints. Your industry partners will be more focused on achieving research outputs which may lead to profitable products; they will have less freedom for exploratory research; they will have less (if any) time to work on academic publications; and they may be more sensitive to intellectual property issues, etc. In this context, it is important to make sure your plans demonstrate relevant awareness of industry culture and practice, ensuring realistic timelines, and project outputs of relevance to their business needs.

Industrial partnerships and intellectual property

Another challenge to effective industrial partnerships can be tensions over intellectual property (IP) ownership. Most research funding agencies have clear IP rules, so if you believe your research may lead to potentially valuable IP then these should be considered carefully as you prepare your grant proposal. In particular, it is important to manage your partner's expectations regarding IP appropriately. Some research agencies require that host universities own all IP generated by grants, while others give universities greater flexibility to negotiate their own agreements.

> *Given the nature of the research and engagement with the industrial partner, the applicant really should have an appropriate intellectual property agreement in place before starting the project.*
>
> **— Quote from a reviewer**

The appropriate terms of an IP agreement may depend on a range of factors: the nature of the research domain, any use of background IP, contributions made by different parties to the research endeavour, etc. Sometimes industrial partners are reluctant to engage in a complex and lengthy process of negotiating the details of an IP agreement related to a project until they know the grant has been awarded. Nevertheless, there are some basic principles related to IP you should consider:

Things that may help convince reviewers:

- **DO** be realistic about the value of IP. Maintaining and defending patents requires significant investment in time and legal fees, without any guarantee of major financial reward.

- If it is not possible to finalize an IP agreement before submitting your proposal, **DO** outline the principles by which such an agreement would be negotiated.

- **DO** follow all IP rules of the funding agency and host university. And make sure your partner is aware of these rules.

- **DO** start working on IP agreements early. These take time. They may need to be approved by company lawyers and the university's knowledge transfer office.

- **DO** get early advice from knowledge transfer professionals in your university.

- **DO** make sure that any collaborative agreements give you appropriate freedom to publish.

Things that may help convince reviewers about your ability to manage your industrial partnership:

- Having someone on the academic team with industry experience.
- Having an industry or 'user' advisory panel.
- Having team members spend time on the industry campus.
- Hosting 'embedded' industry researchers for a period of time.
- Developing appropriate intellectual property arrangements and/or memoranda of understanding, etc.

Challenges associated with international collaboration (and 'collaboration-at-a-distance', more generally)

International partnerships can be very valuable in terms of connecting to new communities and networks, accessing a wider pool of global expertise, and raising your profile. Some funding agencies will have specific grant programmes dedicated to international collaboration or which encourage it.

As the distance (geographical and cultural) increases between partners, so also do the difficulties of managing the relationship.

International Partnership Example: Earth and Atmospheric Science

The right combination of international partners can make the difference between success and failure, especially in research on global challenges.

A proposal to explore and analyse ozone levels over the applicants' country highlighted the paucity of information on ozone distribution in the region, despite increasing global cooperation on ozone monitoring and research. The applicants made a compelling case for the importance of understanding the causes of locally high ozone concentrations (perhaps even high enough to be injurious to human health) given the high concentration of regional industries emitting volatile organic compounds.

Reviewers were especially impressed by the applicants' effort to engage in systematic global ozone monitoring programmes which would help build the country's skills and expertise, as well as its reputation within the global research community. In particular, the reviewers' confidence that the proposal would succeed was *"hugely strengthened by international collaboration with the US National Oceanic and Atmospheric Administration and an impressive range of other international researchers and staff scientists at local airports."*

Even though some modern communication technologies (e.g., Skype and various social networking sites) have made things easier, there are still many difficulties associated with collaboration at a distance. Nothing is a real substitute for face-to-face meetings. It is much easier to establish an effective working relationship and trust in someone when you have had a chance to interact regularly in person and spent some time together getting to know one another.

> *Her plan to send one of her PhD students to [the collaborators] for 3 months and the fact that there will also be one of her partner's students coming in the other direction should ensure effective knowledge transfer and help grow the partnership between the groups.*
>
> **— Quote from a reviewer**

At a distance, there are inevitable difficulties in communicating in a timely and effective way; and a tendency to prioritise tasks involving colleagues you see more regularly. Even chance meetings with colleagues in corridors and common rooms can keep their activities and sub-projects at the forefront of your mind.

Partnership Example: Materials Physics

Movement of researchers between collaborating groups can help underpin a successful partnership

A proposal by an experimental condensed matter physicist which involved carrying out computer simulations of electron dynamics in novel quantum-confined structures was well received by reviewers. This was despite the fact that the applicant had no track record of publications based on computational physics or theoretical modeling research.

The reviewers were impressed by the researcher's plans to send a graduate student to spend three months with a leading computational physics group to learn how to use their code and adapt it for the systems he wanted to study. The reviewers were further convinced by the partnership's *"real synergy"* — in particular, the fact that the collaborating group planned to send a graduate student in the other direction (an *"exchange of hostages"*) to help design structures which would then be studied in real experiments. One group gained a new simulation tool and skills, the other group got to test some of their models and to gain insights into the constraints faced by experimentalists.

In the context of international partnerships, you may also need to consider the implications for 'national impact' (see section on 'National Importance' in Chapter Seven). If your international partner is likely to benefit disproportionately from the partnership this may put off reviewers (or the funding agency), especially if you are engaged in applied, technology-related R&D and there is a risk of too much knowledge flow *out* of your country. Under such circumstances, you may need to work even harder to convince the reviewers of your ability to make the collaboration work and, in particular, that there will be a transfer of valuable knowledge or skills to your country over the course of the collaboration.

Activities that may help convince reviewers:

- Time spent on each other's 'campus'. This is one of the best ways to keep the partnership strong. This needs to be balanced, of course, by the cost of travel and time away from your own lab. If your travel budget starts looking too large, this may also put off the reviewers.
- Time spent by PhD students or post docs with your industry partners.
- Regular scheduled video conferences.
- Plans for working meetings at conferences you will both be attending.

Management and governance

As discussed above, there are a range of challenges associated with making different types of partnership work. Some of these challenges can be overcome by putting more effort into communication, engagement, and hiring team members with the right career experience or personalities. Other challenges can be mitigated by following a couple of partnership management practices:

- Involve your partners in shaping the research project and writing the proposal.
- Plan to put in place a management committee (or other regular forum) for you and your partners to review progress and make joint-decisions.

Involve your partners in writing the proposal

Make sure collaborators have had the opportunity to shape and inform the proposal. Not only will this make the proposal stronger, but it should help you identify and resolve any issues before the grant is awarded. Working on the proposal will also encourage 'buy in' by the partners once the project gets underway.

Make sure any disagreements are identified early. You don't want to discover these just before the submission deadline. This will also start the process of building a working relationship and trust as early as possible, so if your application is successful, it will be off to a better start.

Another reason to engage partners early in the writing process is that they can give valuable feedback on the proposal itself. Not only can they help strengthen those sections of proposal in which they plan to participate, but they can also offer valuable advice as reviewers of the other sections. In particular, input from industrial or social partners can make sections on 'impact' stronger.

If reviewers suspect the proposal has not been informed and shaped by the partners, they may doubt that the partnerships are real. Experienced reviewers will be able to tell if your partners have really engaged.

> *The choice of analysis technique is somewhat surprising... [Technique X] would have been a better choice. Given the collaborators' experience and expertise in X, I can only assume they haven't been closely involved in designing the work package.*
>
> **— Quote from a reviewer**

Management committees and decision-making

Some grant proposal forms will not provide much (if any) space for you to outline any management or governance plans that you have. If, however, your research agenda or partnership arrangements are relatively complex, it may be worth outlining your efforts to manage this complexity. For other grant programmes, e.g., large multi-partner centre grants, the management plans will always be important and are likely to be scrutinized extremely carefully. In any context, it is worth asking yourself the following questions:

- How will your partners be involved in decision-making related to allocating resources or revising research goals and milestones?
- Can your partners' work packages be closed down if they are not working, or if other projects emerge as priorities for investment and effort?
- Are there clear protocols for accessing partners' equipment or for them to access yours?

If you don't have a track record of managing collaborative research projects, then having a clearly thought through management structure and processes for joint decision-making may give the reviewers extra confidence that you will be able to make the partnership work.

> *Given the interdependence of some of the projects and complex agenda, I was pleased to see a reference to the joint management committee. Very sensible.*
>
> **— Quote from a reviewer**

In general, it is a good idea to make sure you have regular meetings with your partners (whether in person or virtually) to update each other on progress and discuss future plans.

If your project is sufficiently large, it may be worth forming an advisory panel which can offer advice based on their expertise both in the design of the project and as it evolves. Such advisors can add credibility to your proposal or even act as 'ambassadors' for your work by raising the profile of your research, by helping to find PhD students or post docs, and by disseminating findings.

Things that may help convince reviewers:

- Evidence that partners have been involved in preparing the proposal, e.g., jointly gathered preliminary data.
- Making sure the partner's contribution is described in sufficient detail.
- Clear plans for regular management meetings.
- Clear process for resolving difficulties, disagreements or disputes.
- Having an advisory panel of experienced researchers who can guide, inform and promote your project.

6.3 Convincing the Reviewers the Partnership is Real

Your research partnerships will, of course, only impress your reviewers if they believe the partnerships are real. In particular, the reviewers will need to be convinced that your collaborators are willing and able to deliver on the tasks assigned to them. There are two main factors which can help make the partnership look credible and substantial:

- A history of successful collaboration or significant engagement.
- Effective letters of commitment and support.

History of engagement and trust

If you have worked successfully with a collaborator before, do stress this in your proposal. It will give the reviewers significant confidence that a productive relationship has already been established and any future collaboration is more likely to be successful and to get up-and-running quickly.

> *The partnership between the applicant and his collaborators in X looks strong. They have worked together before and published some interesting work, which bodes well for the proposed project.*
>
> **— Quote from a reviewer**

As discussed above, even early collaborative efforts to gather preliminary data or other joint activities (e.g., workshops) can give the reviewers added confidence in the seriousness and purpose of your partnership.

If you name a busy 'big name' researcher as a collaborator, this may be greeted with scepticism by reviewers. Important figures in the field tend to be extremely active and get many requests for collaboration. Unless there is a very strong 'letter of support' (see further), reviewers may doubt that such a busy collaborator will be able to devote significant time to the partnership. Sometimes your proposal will be more credible to reviewers if your research partner is less high profile but more likely to have the time and motivation to collaborate in a substantial way.

Effective letters of support

The purpose of letters of support is to persuade the reviewers that the collaborator(s) are appropriate, capable, enthusiastic, and, in particular, **committed** to contributing to the planned research endeavour.

Letters of support can be very important. In particular, a detailed letter from your partner articulating a compelling case for the value of the collaboration **to them** can be very convincing.

- **DO NOT** write a letter of support for your research partners.
- **DO NOT** (under any circumstances) write a sample letter of support for your proposal that your collaborators simply sign and send back to you without adapting it to convey their particular contribution and motivation for engagement.

> *All the letters of support from collaborators are the same. **Exactly** the same. The applicant has clearly written them himself... If the collaborators are not prepared to write their own support letters then I cannot be convinced about their commitment.*
>
> **— Quote from a reviewer**

- **DO NOT** (under any circumstances *whatsoever*) indicate that a researcher will be a collaborator without asking them first. The research community in any particular field can be quite small. There is nothing more embarrassing (or fatal to a grant application) than reviewers somehow being aware that an identified 'collaborator' has not yet been approached.

> *The applicants have cited a colleague of mine as a collaborator, but he tells me has not even been approached, let alone agreed to be a research partner.*
>
> **— Quote from a reviewer**

Statements that you intend to approach a named researcher to be a partner in the future, particularly if you only intend to do this if the funding is successful, are not very convincing for reviewers. Such caution is understandable, but there is no way the reviewers can validate your claims, and they have no way of knowing if the proposed partners would agree to collaborate. Saying you plan to approach famous 'big name' researchers to explore collaboration opportunities are generally greeted with significant scepticism.

> *The role of the international collaborator is unclear. He is obviously well qualified to engage and advise on this project, but the proposal and his 'letter of support' are remarkably unclear about how he will actually assist with the proposed research. WHAT IS HE GOING TO DO?!!!*
>
> **— Quote from a reviewer**

Effective Letters of Support

Convincing letters of support from research partners typically have the following characteristics:

- They are clearly written by your research partner.
- They clearly describe what the collaborator(s) have agreed to contribute to the research endeavour (in terms of expertise, time, access to facilities, etc.)
- They make a compelling case for why the research partnership is valuable to the partner.
- They are printed on institutional letterhead and signed by the partner.
- Where required or appropriate, the letter may also be co-signed by an appropriate institutional party who is authorized on behalf of the partner's university to make any commitments of support.
- They are addressed to you, as the lead principal investigator (PI).
- They address any issues or guidelines identified in the Funding Agency's CFP.

The very substantive and carefully detailed letters of support from the international collaborators indicate there is already good engagement between the researchers and significant commitment.

— Quote from a reviewer

Example of a weak 'Letter of Support'

Letters of Support which only express general interest in the proposed project without **commitment, enthusiasm and details** about what the partner will contribute are worth very little and may even leave a negative impression with the reviewers.

ACME Corporation

Dear Professor B. O'Ffin,

Thank you for your email about your project related to XYZ.

Research in this area is relevant to our company. I look forward to hearing more about your project and its results.

Yours sincerely,

Ivor Company

6.4 Summary

- **DO** clearly identify what your partners are going to contribute to the proposed research agenda. Give details. Explain how their contribution is going to help the project deliver its goals.
- **DO** get substantive and detailed letters of support from partners.
- **DO NOT** add 'big name' researchers to your list of partners, if they are really only acting as advisors.
- Where appropriate, **DO** outline how you will manage the partnership — i.e., processes for resolving conflicts, prioritising resources and revising project goals.
- Where possible, **DO** give evidence of a successful track record of interaction and collaboration (and joint outputs).
- If your proposal highlights future impact in a particular sector or social domain, **DO** include practitioner partners or advisors.
- **DO** involve your partners in writing the proposal and shaping the research agenda. Do this as early as possible.
- **DO** make clear why the work to be done by partners is critical to the overall goals of the project and is not 'tacked on'.
- **DO** make sure your collaborator's resume highlights the areas of knowledge and capability relevant to your proposed joint activity.

- **DO** highlight any evidence of experience in participating in (or, even better, managing) collaborative research projects.
- **DO NOT** overstate the time commitment of your partners.
- If your partners are not local, **DO** describe the ways you will ensure effective communication and knowledge exchange.
- Budget permitting, **DO** plan for team members to spend time with your partner's group.

Chapter Seven

Impact

There cannot be a greater mistake than that of looking superciliously upon practical applications of science. The life and soul of science is its practical application.

— Lord Kelvin

7.1 Introduction

The discovery of the first antibiotic, penicillin, by Alexander Fleming in 1928 dramatically changed the practice of medicine and improved health outcomes for millions of people. Only very few of us can hope to see such an immediate, dramatic impact from our research. However, it is always valuable to keep in mind that improving people's lives in some way should be one important outcome from any research project. In addition to considering the impact of proposed research on the academic world — that is advancing knowledge within particular research fields — many research funding organisations will also consider the broader potential impacts on society and the economy.

Many national research foundations have been created to, at least in part, stimulate economic productivity and growth. Consequently, they need to be able to answer to government (and taxpayers) about the difference their portfolio of investments is making, not only to academia but also to the wellbeing of the country's citizens and the nation's economy. Researchers applying to these funders may be asked to explain the value of their research in terms of national industrial or socio-economic impact.

111

Communicating the socio-economic impact of your research to expert technical reviewers (and especially international reviewers) can be extremely challenging. This chapter shares lessons and practices for effectively communicating the potential 'impact' of research within a grant proposal. Although focusing on industrial and economic impact, this chapter addresses a broad spectrum of types of impact, including:

- **Industrial and economic competitiveness**: Research generating new knowledge that may underpin new products or processes, generate valuable intellectual property, act as a platform for new companies, attract inward investment, etc.
- **Skills, talent and capabilities**: Research activities that enhance the skills, capabilities and training of the next generation of research talent or the skills and capabilities of industry partners.
- **Societal health, wellbeing and culture**: Research addressing important societal challenges (e.g., climate change, healthcare, defence, mobility, etc). research addressing the evidence needs of public sector organisations and policy-makers; or research outputs that may enhance national cultural life.

Different research fields and different types of project may, of course, have very different outputs, outcomes, and beneficiaries, as well as different time frames and routes to impact. Although most national research funding agencies do not have fixed metrics or rules for assessing impact within their decision-making process, the quality of the impact 'story' in a research proposal can sometimes make the difference between a proposal being funded or not. This may be particularly important in research competitions focused on industrial partnerships or technology transfer. Remember, however, you can help convince reviewers about the potential impact of your research by:

- Demonstrating the ability to describe plausible '**pathways to impact**' from the knowledge generated in your research to changes in industry or society that will be effected by this new knowledge.
- Drawing upon **user insights** from practitioners in the domain you hope your research will have an impact on.

- Having in place plans, processes or partnerships to ensure your **findings are disseminated** to the right people who can take them on the next stage of the journey along the 'pathway to impact'.

This chapter addresses a range of themes and issues which are important to communicate the impact of a proposed research project convincingly, including:

- 'Pathways' and timelines to impact: *"How and when will this matter?"*
  - Outputs, outcomes and impact evaluation.
  - Local, national and global impact.
- Users and beneficiaries: *"Who cares?"*
  - Economy and Industry.
  - Society and Public Sector.
  - Skills and education.
- Dissemination and foresight: *"Who's next? What happens next?"*
  - Who will take your findings to the next stage of development?
  - Anticipating future challenges along the 'pathway to impact'.

7.2 Pathways and Timelines to Impact

As discussed above, research projects can have very different types of outputs, beneficiaries and time frames for impact. Apart from very specific industry-focussed funding competitions, research funding agencies typically try to balance a portfolio of projects and have relatively flexible methods for assessing impact. Consequently, there is no one-size-fits-all 'correct way' of making the case for the impact of a proposed research project.

'Pathways to impact'

When making the case for the future impact of a research project, it is important to offer more than just a vision for how things might be different in the future because of your research. In particular, it is important to be able to make the case for the 'pathway to impact' — how the new knowledge generated by your work will be translated into economy or society — and, in particular, who will take the knowledge generated by your research to the next stage on this journey.

Some researchers make the mistake of only focusing on illustrations of how their research (or research from their field) has made an impact in the past. A track record of success is important and can help make the case for your proposed research, but the main emphasis of your discussion of impact should be primarily **forward-looking**.

Therefore, one of the most important aspects of communicating the future impact of a research project is to have a compelling narrative for this pathway (or pathways) — how the knowledge generated by the proposed research may eventually benefit particular users, and what other partners or stakeholders may need to be involved at different stages of the journey.

> *The applicant makes a good (if somewhat verbose) case that this emerging technology domain has great promise and the market could be very large. What he does not do is make a sufficiently compelling case for the particular contribution and **impact of this specific project**.*
>
> **— Quote from reviewer**

The 'pathway to impact' in an emerging or smaller national innovation system can be much more challenging than in more established, larger ones. In particular, there is a greater likelihood that not all the necessary expertise, infrastructure or organizations will exist within the country to help translate new research knowledge into real economic or social impact. Under these circumstances, reviewers may be sceptical about the prospects of the research creating significant value in the country.

> *The close engagement with the partner company should significantly increase the probability that the outputs will end up in useful technologies.*
>
> *It is a pity, however, that the partner company has no industrial operations in the country. In this reviewer's opinion, this means there is a lower probability of significant economic impact from this proposal relative to some others in this competition.*
>
> **— Quote from reviewer**

As part of this impact narrative, it is important to describe your plans to increase the efficiency of your impact efforts and to enhance the likelihood that the research outputs will be taken up by developers and users. This should, in turn, accelerate the timeline to impact and optimize the benefits of your research. In this context, it is important to identify the ways the research agenda will be informed by potential users and beneficiaries, in particular by indicating what other researchers, developers or professionals will need to be involved at different stages, and by outlining your strategy for identifying and engaging with them.

Timelines to impact

There can be significant variations between projects in terms of the duration between the initial outputs to when this new knowledge starts to make a significant difference in the 'real world'. Sometimes impact can be instantaneous; sometimes the full impact of research findings may take decades. These timelines depend on a variety of factors, including: the maturity of the research field; how applied the project is; the openness of users to new ideas; and the extent to which alternative theories, technologies or practices are deeply embedded. For example, a new computer algorithm for searching the web may have a much shorter 'time to market' than the discovery of a new pathway to treat a disease. Again, most research funding agencies will not have fixed rules or expectations about what an acceptable time to impact is. If your suggested time to impact is very long and if many other researchers' efforts will subsequently be necessary for real impact to be achieved, reviewers may find your impact statement less compelling than others.

Isaac Newton struggles to write the economic impact section of his 'gravity' proposal.

Things that may help convince reviewers:

- **DO** give realistic timelines for the impact of your research. If the reviewers do not find your timescale plausible this may undermine their faith in your ability to contribute to impact at all.

- **DO NOT** oversell the impact of your proposed research. If future impact will also rely on further contributions by other researchers and stakeholders, do not imply impact attribution to your work alone.

Outputs, outcomes and impact evaluation

When writing about the economic or social relevance of your research in your grant application one of the most important things to bear in mind is that **'outputs' are not the same as 'outcomes'** and some agencies even draw a distinction between 'outcomes' and 'impact'.

Research outputs are generally understood as the direct, measurable **deliverables** from your research activities. Outcomes are **changes** that result from these outputs being used in the academic or the real world. The words 'outcomes' and 'impact' are often used synonymously, but some funding agencies use terms like 'economic impact' or 'social impact' to mean the overall **benefits** to society or the economy that follow from the outcomes. Some of these distinctions are summarized and illustrated in Table 7.1.

Different funding agencies in different countries may use terms like 'outcomes' and 'impact' slightly differently. When writing your grant application you should, of course, use terms in the same way as the funding agency. Nevertheless, when telling your story about the relevance of your research it is useful not only to distinguish between **deliverables, changes and benefits,** but also to make a compelling case for how your outputs may be turned into outcomes and how these may evolve into real economic or social benefits. A clear 'pathways to impact' story will strengthen your proposal.

Table 7.1: Outputs, outcomes and impact.

Outputs	The **measurable deliverables** or results of your research activity	• journal articles • conference papers • patents • students graduated
Outcomes	The **changes** that result from your research outputs being used	• new knowledge transferred to users • new technologies • new companies • new members of the workforce
Impact (Economic, Societal or Cultural)	The overall **benefit** to society or the economy caused by the outcomes of your research	• longer/better quality of life • improved competitiveness of national firms in particular sectors • enhanced public services • increased employment opportunities

> *A patent is an output not an outcome ... A new technology based on the patent is an outcome but this still won't have any real impact on the economy until that technology is in an actual product and generating revenue.*
>
> **— Quote from reviewer**

Things that may help convince reviewers:

- **DO** provide evidence of a track record of impact. Make sure you describe previous research where there has already been a successful impact.

- **DO** put plans in place to evaluate the impact of the project. This can help convince reviewers you have your 'eyes on the prize' of impact. If your plans appear appropriate, this may increase their confidence in your judgement and ambition.

- **DO NOT** over claim. Any impact from research on the real world is likely to be ultimately built on a collective effort and community-wide accumulation of knowledge and infrastructure.

National importance

Many national research funding agencies are putting increasing emphasis on the **national** impact of research projects — i.e., impact on users and beneficiaries within the national economy or society, as opposed to research which may impact particular industries globally. Under these circumstances, applicants should make as compelling a case as possible for how their planned research outputs will result in value for their country.

> *All of Professor X's industry partners are based outside the country. It seems a shame that (even if this project is successful) the value from the next phases of development, manufacturing and exploitation will most likely be captured elsewhere.*
>
> **— Quote from reviewer**

Many governments in smaller countries or nations with emerging innovation systems will have formulated National Development Plans or strategies for priority sectors. These documents (as well as any underpinning economic analyses or benchmarking studies) can be a very useful resource for connecting the outputs and outcomes of your research to key competencies, industrial strengths and sectors of importance to your country.

As discussed later in this chapter, the impact of your research may be on your nation's society rather than industry or economy. For example, life science or medical research may directly improve the health and well-being of national citizens. In some cases this may be highly specific to your country, for example tackling diseases which are especially prevalent within the national population or addressing indigenous medical conditions that may have not received the same attention as diseases prevalent in countries with more established medical research systems.

> *The proposal does not address the Research Foundation's expectations regarding* **specific** *benefits to the* **national** *economy. The international industrial partner will get to test its next generation machine within the national medical research system — and local researchers may benefit from using leading-edge imaging technology — but once the testing is over there is no indication that the industrial partner has any plans to carry out R&D or manufacturing. I can see no tangible* **economic** *impact to be left behind.*
>
> **— Quote from reviewer**

Furthermore, many high-growth emerging economies are experiencing serious social 'grand challenges', for example: Rapid urbanization, high levels of pollution, limited access to natural mineral resources, energy, and water. Again, these challenges are likely to have been studied by government agencies or national institutes. Drawing upon such analyses to make the case for the value of your research to society and the importance of its impact can be very effective.

Impact Case Example: Medical Engineering

Real economic impact can come from strengthening the critical mass and international reputation of a country's research expertise.

A multidisciplinary (biomedical engineering/computer science) proposal aimed to develop a new medical tool which would provide implant surgeons with real-time force feedback data, together with patient-specific imaging of the location of the tool.

Reviewers thought that the proposed application in the domain of cochlear implantation surgery was particularly relevant and appropriate; and that improved success in this area would not only have direct medical and health impacts, but would also add to complementary research activities in the country, leading to its becoming "a worldwide beacon of excellence in the area" with implications for potential economic impact as well.

Things that may help convince reviewers

- **DO** quote any relevant governmental studies (or reports by other national institutions) which identify the importance of advancing knowledge in areas addressed by your proposed research.

- **DO**, where appropriate, engage with national firms (or important local operations of international firms) or institutions.

- **DO** highlight any evidence that your research will add to a critical mass of research excellence in a domain of importance to the national economy or society.

- **DO** explain how your research is consistent with (or complementary to) other national research and innovation investments or investment priorities.

7.3 Users and Beneficiaries: "Who Cares? Who Benefits?"

A critical part of communicating the potential impact of a research project is demonstrating awareness of the relevant user community and identifying where and when it may be necessary to engage with them. The ultimate impact of a research project can be significantly enhanced by paying attention to the following activities:

- Building trust and engaging with users of the research outputs (where possible collaborating with established networks of user groups).
- Involving users at all stages of the research process from project design, to review and dissemination.
- Working with developers and users to anticipate barriers to the take-up of research outputs (and ultimate impact) and identifying strategies for overcoming these challenges.

> *The applicant has clearly worked closely with his industrial advisors to anticipate challenges this tele-communications device may face in later stages of its development and system integration ... he has built these insights into his research strategy, e.g., aware-ness of likely industry standards requirements has influenced choices of materials used in experiments, increasing their relevance and likely impact.*
>
> **— Quote from reviewer**

Research impact: Who? What? How?

In particular, research proposals should, where appropriate, clearly iden-tify and describe the following:

- Who the ultimate beneficiaries are likely to be.
- How they will benefit.
- Other beneficiaries/stakeholders who may have an interest in outcomes.
- Other researchers, developers or professionals who will need to be involved in the planned and future research efforts to ensure impact.
- How you are already engaged with (or plan to engage with) these groups.
- How the work will build on any specific existing links.

Impact Case Example: Civil Engineering and Earth Science

Even a proposal with strong cases for potential social *and* economic impact can be undermined by inadequate engagement with practitioners.

A proposal to study seismic risk in a country with a distinct geology and dramatic growth in high-rise building infrastructure made a compelling case for its potential to both save lives and improve the knowledge and skills of the local building industry.

Reviewers thought the team was strong and technical approach appropriate and that the project had the potential, *in principle*, of generating knowledge that would be *invaluable for the country*.

Nevertheless, reviewers criticized the applicants for not strengthening their 'pathways to impact' story by engaging closely with relevant government authorities such as the Ministry of Public Infrastructure and Housing and by collaborating directly with local construction firms.

Some practices for effectively communicating the nature, quality and added value of research partnerships (and convincing reviewers of their importance) are outlined in Chapter Six. As with research collaboration partners, proposals which build on established relationships with user partners, in particular where there is evidence of successful engagement and impact in the past, are especially compelling.

If at all possible, involving potential users (and beneficiaries) in informing the development of a research project and obtaining **substantial** letters of support or commitment prior to submitting a funding request can significantly enhance a proposal.

In addition to advising on how the proposed research agenda will be grounded in the real world needs of the intended beneficiaries (and giving reviewers added confidence in this regard), users may also be able to advise on the most effective ways of disseminating research findings.

Things that may help convince reviewers:

- Evidence that you have engaged with potential users of your research outputs in the design of the proposal.

- Where possible and appropriate, collaboration with established stakeholder organizations (or relevant user networks).

- Explicitly including in your research strategy activities to engage with users to help anticipate barriers to the take-up of your planned research outputs (and ultimate impact); and identifying strategies for overcoming these challenges.

People: Education, skills and broadening participation

Normally, any grant application sections related to 'impact' are intended to be used to describe the benefits of your research **beyond normal academic or teaching audiences**, in particular the benefit to industry, the public sector, charities, or even the wider public. Nevertheless, academic or teaching activities directly related to your proposed research project may contribute to enabling ultimate economic and societal impact and may, therefore, be appropriate to include under these circumstances.

> *I suspect that the most important impact of this work — even more than influence of the proposed outputs — will be training of the industry-relevant skilled post-doctoral researchers in this important emerging technology domain.*
>
> **— Quote from a reviewer**

Education-related activities that might be especially worth including in any discussion of impact include:

- Training of graduate students (or even advanced undergraduates) in the use of cutting-edge techniques and equipment, in particular in emerging technology related areas of relevance to industry.

- Specialist training for industry research partners (or industry professionals more generally) related to new knowledge, techniques or practices generated by the proposed research project, especially when this is relevant to important industrial or societal challenges.
- Training for other public sector professionals, such as healthcare professionals or policy makers.
- The development of course modules or materials based on new knowledge or techniques. Evidence that such materials may be disseminated and adopted more broadly beyond the host institution can be especially compelling.

Some funding agencies will also take into account how the proposed research might positively impact the composition of national research communities in terms of broadening the participation of underrepresented groups (e.g., women in certain disciplines, ethnic minorities, people with disabilities, etc.) If your research activities are likely to have a significant impact in any of these ways, this may be worth highlighting.

Things that may help convince reviewers:

- **DO NOT** include in your impact statement any teaching activities that might be considered a normal part of the duties of an academic. Any education related activities you cite should clearly be clearly linked to the proposed project and should make an impact beyond that which might normally be expected.
- **DO** include statements from relevant stakeholders (e.g., employers) about the value of the enhanced curriculum and likely impact on their activities and capabilities.

Impact on the economy and private sector

As noted in Section 7.1 above, you should take care not to confuse outputs with outcomes. Describing different metrics (e.g., numbers of patents, numbers of students trained for the workforce, etc). is not the same as describing the impact you anticipate these outputs having on the real world. However, these output goals are worth including in your proposal. Sensible targets will give reviewers a sense of your planning abilities, ambition and focus. But they are not the same as impact.

A patent, for example, is not an impact. Outputs like patents only have an impact **if they are actually used**. They will start to seriously impact the national economy only when they underpin real technologies, generate revenue and give competitive advantage to real firms that employ real people. Your case for impact will be more impressive if you can make a compelling case for how the knowledge you generate will be used to develop new technologies and products.

> *The track record of the PI's collaborators in taking new technologies to market (albeit in a different application domain) should increase the likelihood of translation and commercialization.*
>
> **— Quote from a reviewer**

Similarly, educating PhD students can, of course, be of value to industry, but only if they are hired by companies. If you can make a strong case that you have a track record of producing PhDs that go on to work in industry, and that you are equipping your students with key skills to help them play a valuable role in the future of those industries, the case for impact will be stronger.

It is important to remember that there are a wide variety of types of private sector actors who may benefit from your research outputs or help translate them into economic impact.

Efforts made to align the research agenda with the industry association 'technology roadmap' should increase alignment with private sector investments and the likelihood of take up by industry.

Examples of private sector users include:

- R&D-intensive multinational firms
- manufacturing operations of major firms
- small and medium-sized enterprises
- start-ups
- industry associations
- industrial research consortia
- standards development bodies

Furthermore, there may be impact across the entire industrial 'value chain', i.e., not only bringing new knowledge to private sector research endeavours, but influencing design, manufacturing, logistics, operations management, etc. Or your research may benefit more than one industrial sector.

> *The applicants' [production engineering] research is likely to be of value to industry, but to ensure their findings are as relevant as possible, they should also engage with the manufacturing divisions of partner firms.*
>
> **— Quote from a reviewer**

Impact Case Study: Management Science and Marketing

Projects with high economic impact potential but which use relatively established techniques may look more like consulting than research.

A project focused on improving the quality of tourism services was well received by reviewers in terms of its potential impact on an emerging economy where tourism is a major source of national revenue. The applicants proposed using a well established model for quantitatively exploring and assessing customers' service experiences, attitudes and perceptions. The reviewers thought the project was *highly relevant in terms of contributing to economic development.*

The reviewers also thought, however, that the proposal was *not really a piece of research that will expand the frontiers of knowledge. Rather it is a piece of academic consultancy.* The reviewers believed that the data gathered would indeed improve tourism policy making as well as the practices of tourism organisations. The fact that the economic impact case was so compelling — combined with the fact that the methods to be used were so well established — led some of the reviewers to believe that the project *could and should be more appropriately funded by the tourism industry itself.*

Things that may help convince reviewers:

- Alignment of research goals with industry roadmaps.
- Evidence of having engaged with industry partners or industry associations in shaping your research goals.
- A track record of engagement with corporate researchers and seeing the translation of knowledge generated by your projects into technologies, manufacturing processes, products or services.
- Team members with relevant industry experience, who can inform the research agenda with insights into how research findings can have a real impact.

Impact on society

Although many agencies that fund science and engineering research can be particularly focused on impact on industry and the economy (especially in terms of advances in technology, manufacturing and product innovation), most will also value impact on important societal challenges facing the health, well-being, or even cultural life, of a nation's citizens.

For this reason it is worth considering the different types of public sector, charitable or other not-for-profit actors who may benefit from your research or help translate your outputs into social impact. It may be helpful to identify private sector actors whose new products or services will improve people's lives (as well as company balance sheets).

Examples of public sector or not-for-profit users include:

- hospitals
- policy-makers
- regulators
- charities
- government agencies
- schools
- learned societies
- professional bodies
- museums
- performing arts organizations

In particular, there is a range of research 'grand challenges' related to key issues facing modern societies, for example: health and well-being, transport and mobility, energy, climate change and the environment, aging population, urbanization and the cities of the future. Research outputs which may make a significant contribution to tackling challenges in these areas should also be highlighted when discussing the impact of your proposed project.

Things that may help convince reviewers:

- Alignment of research goals with the strategies of relevant public sector agencies or societal groups.
- Evidence of having engaged with stakeholders in shaping research goals and planned outputs to optimize impact.
- A history of translating knowledge from your research into non-profit activities and having an impact on societal challenges.
- Team members with relevant career experience in relevant organizations (government, healthcare, charities, etc.), who can inform the research agenda with insights into how research findings can have a real impact.

7.4 Knowledge Exchange and Dissemination: "Who's Next?"

The ability to disseminate the new knowledge (including know-how and skills) generated by your research may be critical to its ultimate impact. It is can be helpful to describe any dissemination efforts or knowledge exchange activities with users or beneficiaries that you have planned.

Knowledge exchange activities

Some activities which may be useful in ensuring effective engagement and knowledge transfer to users and beneficiaries include:

- Seconding team members to user organizations (or hosting inward secondments of user community staff).

- Training workshops involving user groups.
- Tutorials at conferences.
- Publications in industry magazines or other user-oriented media summarizing main outcomes in an accessible and useable way.
- Research planning workshops which involve potential users in order to better understand what information and insights they need and what format your research outputs should take to be as accessible and helpful as possible.
- Freeware tools.
- Open online databases (especially if your research has generated large amounts of data which may be useful for users, if shared in an accessible format).
- 'Workbooks' describing new techniques or methodologies, sometimes in video format.

Dissemination plans

If the grant application process requests information about impact, it is generally worth including some details about any plans you have for disseminating your research to user groups or other beneficiaries. One thing that can make your dissemination plans compelling to the reviewers is the experience and skills of those involved in dissemination. Sometimes research teams can effectively engage specialist staff to undertake communication and exploitation activities, or they can engage technical communication experts to write publications and web pages which are accessible to the target user groups and beneficiaries.

> *The private sector experience of the 'industrial outreach officer' is impressive and likely to significantly enhance dissemination to users ... and, ultimately, impact.*
>
> **— Quote from a reviewer**

If your existing team members have relevant knowledge exchange skills and track records of successful translation of new knowledge to users, it is worth highlighting these facts. If there is no track record of dissemination or if relevant skills are missing, however, it may be worth identifying how you plan to address this.

Things that may help convince reviewers

- **DO** provide any evidence of having engaged with user groups or beneficiary organizations shaping the dissemination plan.

- **DO** plan for ongoing engagement with user groups to understand how to communicate new findings as they develop.

- **DO** highlight any track record of engagement with beneficiaries and successful dissemination of research findings to user groups.

- **DO** outline clear and appropriately detailed plans for dissemination.

- Where possible, **DO** involve team members with a background in the user community you hope to impact.

7.4 Summary

- **DO NOT** focus too much on past outcomes. A track record of success is important, but the case for the impact of your proposed research should be primarily forward-looking. Nevertheless, mentioning past impacts of your research can be helpful.

- **DO** clearly identify users and beneficiaries, and *how* they will benefit.

- When making the case for the impact of your work, **DO NOT** just tell a story about how industry or society may be different in the future because of your research. It is important to be able to make the case for the 'pathway to impact' — how the new knowledge generated by your work will be translated into the economy or society and, where appropriate, who will take the knowledge generated by your research to the next stage on its journey along this pathway.

- **DO** describe any planned dissemination efforts or knowledge exchange activities with users or beneficiaries. This will give reviewers greater confidence in the likelihood of your research work's potential impact being achieved.

- **DO** give information about who will be engaged in knowledge exchange activities, including any relevant expertise or career experience that would enhance the effectiveness of communication with users and other beneficiaries of your research.

- Remember that for many funding agencies, impact is not just about economic impact and knowledge transfer to firms. **DO** highlight any impact which can make a difference to society, including the health and wellbeing of people as well as contributions to the quality of the nation's cultural life.

- Where appropriate, **DO** distinguish between the impact of your proposed research on your local national economy and society, and the wider impacts on the global economy and wellbeing of people generally. The extent to which the value of your research can be captured by your nation may be an important final consideration in the decision to fund your research.

Chapter Eight

Referencing, Plagiarism and Intellectual Property

Great discoveries and improvements invariably involve the cooperation of many minds.

— Alexander Graham Bell

So far, we have dealt with generating the idea for a proposal, producing a first draft and the importance and mechanics of finalizing this draft. We have also discussed making the case for partners, if needed, plus the impact of the project. We now turn to other aspects of grant-writing which can be very important professionally and which can have a large effect on the success of your proposal. These include proper referencing to other relevant work, honesty about assigning credit, and dealing with the intellectual property implications of your project.

8.1 Referencing

Credit for an idea is the essential currency of science, and most scientists are scrupulous about giving and claiming credit for particular ideas. This makes it all the more important to give proper credit to the originator of any idea in your proposal.

> *There are only two publications cited, both about seven years old. While the authors state that they have published related work to that listed in the proposal, they do not provide detailed references to this work. Given the large amount of literature in this area, the lack of a more extensive list of references is a severe weakness of the proposal.*
>
> **— Quote from a reviewer**

Referring to earlier work, either your own or others, is also very useful in this context. It allows you to **provide additional information about the background of the project without** taking up a tremendous amount of space in your proposal. You can use references to contextualize your proposal within the current state of the field and to show both how it connects to past work and how it will move the field forward.

Another reason to reference earlier work is to anticipate questions that reviewers might raise about either your assumptions or about the methodology that you plan to use (see also Section 4.2). If these methods have been used successfully previously, then **referencing the earlier work provides good evidence of the validity of your approach.** This can be especially powerful if the earlier published work was carried out by reputable scientists and is well established. Any of your own work related to your new proposal that is already published also provides evidence that your current project is likely to work using the same, or at least very similar, techniques.

> *The list of references is excellent. These are referred to appropriately in the body of the proposal and are used to motivate and explain the work proposed. These references are all in up to date, well respected, international journals indicating that the authors are familiar with the current work in the field.*
>
> **— Quote from a reviewer**

If the technology of your project is quite complicated, or you are using a complex (and expensive) piece of equipment, references to earlier work using this same equipment can save you space in your new proposal. Just be sure that the reference you choose to provide does indeed supply sufficient information about the use of the equipment to satisfy any questions the reviewer might have. Reviewers will not necessarily look up and read all the references you provide, but they may.

You should also provide a sufficient number of relevant references **to indicate that you are very familiar with the current state of the field** of your proposal. Even if you have not yet published extensively in the field, you can show your knowledge of the issues by providing appropriate references. If your proposal is going to advance the field, it is important to show that you are familiar with the important current issues.

> *The references given are mainly to the work of the current group and they are quite dated. The most recent reference is dated 1997. This leads me to question how current the present proposers are in their work.*
>
> **— Quote from a reviewer**

In most fields of science and technology, a relatively few journals are normally the most important ones. Reviewers in the different scientific fields will be aware of which journals these are. In many cases 'letter' journals which reflect new and important breakthroughs are seen as most important. In other cases, review journals with articles summarizing a discrete area of science are viewed as very prestigious.

One measure of the importance of a particular journal that is often used is the **impact factor**. **This refers to the number of citations per paper published** in the journal. The more citations per paper the better and the more highly the publication is rated. The top ten journals with the highest impact factor are all in biological and medical sciences, reflecting the large number of researchers in these areas. Journals in various fields with high impact factors include *Nature*, *Science*, *Cell* and *Physical Review Letters*. The impact factor is a guide to the importance of a journal, and you should be familiar with the published work in the appropriate important journals in your own field and refer to them in preparing your proposal. You should not rely mainly, and **certainly not exclusively**, on local journals or unpublished reports in your reference list.

Should you also provide references to your own work? Yes, provided the work is relevant to the proposal and, provided that it is published in a reputable international journal. Such references will help convince the reviewers of your familiarity with the field, as well as your productivity, and give them confidence that you can carry out the work proposed.

It would, however, be a mistake to reference only your own work. Certainly, don't put in contrived/tangential references to your own work in an effort to show that you are well published. That is what your CV is for. Instead, you should give credit to other researchers as well as yourself.

On a practical note, make sure that the references are given in sufficient detail that they can be readily found by the reviewers. Reviewers may or may not read the references in your proposal but if they do elect to look them up, any errors or omissions in the citations provided will suggest carelessness on the part of the author. This is not a winning characteristic for the author of a competitive proposal. **The worst error of all is to provide an incorrect reference** or one that does not make the point you claim or that is not relevant to the proposal. If reviewers do choose to look up your references and find some of these problems, they will certainly decrease the score for your proposal. All of these problems fit into the category of **'do not annoy your reviewers'**!

> *The list of references presented, while generally relevant and current was incomplete. Also, Reference Three was misquoted. Instead of Physical Review, I believe the correct reference is to Physical Review Letters. In addition, there are other incomplete and incorrect references given. All this suggests that the reference list could use some careful editing to allow the reviewer to have access to the material without so much searching.*
>
> **— Quote from a reviewer**

We have also seen situations where an author, in describing the tasks to be carried out if the proposal were to be funded, lists a literature survey as the first task, sometimes taking as long as many months. As discussed in Section 4.2, most, if not all, reviewers would look on this very negatively. **The time to carry out a literature survey is before the proposal is submitted** and the results of such a survey should be obvious in the references cited in the proposal.

> *The first year is to be spent defining "work to be undertaken"(?!) ... The applicant should know this already!!!*
>
> *A literature survey is not a 'major' deliverable.*
>
> **— Quotes from reviewers**

8.2 Plagiarism

Plagiarism is copying someone else's work without proper attribution. This is something that is increasingly easy to do in this age of electronic availability of published material. However, if discovered in a proposal, directly copying material without giving credit to the original authors will not only result in the proposal being returned without review (and of course without being funded) but it may also result in academic sanctions against the author(s) of the proposal. **Plagiarism is regarded as a major academic sin**.

While the availability of published material on the internet and the ease of copying such material into other documents makes it tempting to simply copy and paste sections of other documents into a proposal, the use of computer searches also makes such direct copying easier to find. Professors in universities have grappled with this issue with student papers for many years and now the ability to find unattributed material copied from other sources is becoming easier and easier.

It can also happen that one of the reviewers of a proposal, who after all is normally familiar with the field, will recognize a paragraph or page which has been directly copied. This happened to a proposal being reviewed at a funding agency with which we are familiar. One of the reviewers turned out to be the author of a paper from which several paragraphs had been copied word for word. At the panel meeting of all reviewers, this reviewer pointed out the copied paragraphs and the proposal was summarily rejected by the panel without further review. The panel also recommended that the agency contact the authors' university and seek an explanation and, if need be, impose sanctions on the authors. The senior author of the proposal had allocated this section to a postdoctoral fellow who had tried to speed up the process by copying from other published work without giving credit. The senior author had been unaware of this but was nevertheless banned from submitting proposals to that funding agency for a period of time. **The lesson is that plagiarism is very serious and could result in very severe consequences**.

Remember that this is not to say that the results and ideas contained in other published work cannot be used. Far from it! This is how science progresses. But if you wish to quote directly from other published work, then you should surround the material by quotation marks to show that you are using material which has been directly copied from another publication, and of course you should use appropriate citations. It is not acceptable to simply copy from other documents unless quotation marks are used and certainly not without giving proper credit by referencing the original work.

While proper quotations and citations may not be the common practice in all cultures, in science research, it is the absolute norm. Failure to adhere to this norm can have grave consequences. As mentioned earlier in this chapter, **proper credit for ideas (and language) is extremely important to scientists** and not assigning appropriate credit for other people's work will not be readily forgiven.

DO make sure that your fellow writers are familiar with the rules regarding crediting other people's work.

DO NOT use material from other publications without using quotation marks and giving proper references to the work.

8.3 Intellectual Property

One of the main motivations for government support of research, especially science and engineering research, is to move their economies 'up the value chain', in particular to develop national capabilities to carry out higher value added strategy and technology (S&T)-based industrial activities. Often countries see this outcome arising from collaboration between academia and industry, so the funding for research is often linked to such collaborative endeavors. These links can be encouraged by requiring matching funding from industry of some form, either cash or so-called 'in-kind' funding (i.e., through providing access to valuable equipment, staff time or services). Many large industries welcome collaborations with scientists working at the forefront of research, since this gives them some guidance for their own long term projects. These collaborations also provide the opportunity for recruitment of highly qualified junior researchers into the company.

However **collaborations between academia or pure research facilities and industry are often hampered by issues related to intellectual property**. Questions will arise as to who owns the knowledge, the ideas or the technology that has been produced by the joint project. The resolution of these questions can take considerable time and effort and can prove a major impediment to collaborations.

Normally it is wise to try to reach a signed agreement about intellectual property ownership before the proposal is submitted and certainly before the work is carried out. Once some interesting (and potentially valuable) technology has been produced, the stakes become higher so that it becomes more difficult to reach agreement on who should benefit from the discovery. Therefore, an intellectual property agreement should be discussed, and if possible signed, between the parties as soon as possible.

In practice, however this often turns out to be quite difficult. Universities are often suspicious of the motives of industry and industry naturally is trying to maximize the return on its investment. In addition, legal teams from both sides are often involved, especially for large or complex projects. Lawyers have a different culture and different function from the research participants involved in the collaboration. Part of their job is to ensure their side is not taking any undue risks and, consequently, their careful analysis often leads to long delays in reaching agreements. These delays can slow down the start or even the submission of a collaborative proposal between academia and industry.

In an effort to overcome these difficulties and expedite such collaborations, some countries, including Ireland, have attempted to formulate a universal intellectual property agreement for all of their funding agencies. In the case of Ireland, the process of developing a universal agreement took many years and remained controversial for some time.

In general, such universal agreements have had mixed success but often make a good starting point for negotiations in specific cases. The lesson for the author of a proposal which involves cooperation with industry is to **begin to deal with intellectual property questions early in the collaboration**.

In 2012, an 84-page document[1] was released by the Irish government entitled *'Intellectual Property Protocol — Putting public research to work for Ireland: Policies and procedures to help industry make good use of Ireland's public research institutions'*. A brief summary of the Principles stated in this document is given in the following block. These principles emphasize the importance of training research staff to be aware of IP issues. They also provide a **useful guide to principles that are relevant for countries interested in fostering strong ties between university research and industry**.

4.1 *Principles applicable to all forms of research*

4.1.1 IP identification

Research Producing Organizations (RPOs) shall have procedures in place to identify in a timely manner all Intellectual Property (IP) arising from their research. They shall, together with their Technology Transfer Offices (TTOs), support their researchers to help them recognize when their discoveries may have commercial value.

RPOs should work together to identify IP created by different RPOs which, when brought together into a single package, may have commercial value.

(Continued)

(Continued)

4.1.2 IP protection

RPOs shall make clear to their staff, contractors, consultants and students their responsibilities in relation to the protection of IP including the maintenance of research laboratory records and the prevention of premature public disclosure of IP. RPOs shall as far as possible help their staff, contractors, consultants and students to meet these responsibilities.

4.1.3 IP ownership

The ownership of IP arising from research performed by RPOs shall at all times be made clear and unambiguous. RPOs shall have in place, and enforce, arrangements to ensure that initial ownership of IP arising from their research is clearly and unambiguously defined. In particular, RPOs shall ensure that all employees, and non-employees such as contractors, consultants and students, assign to the RPO all rights to IP arising from their research for or on behalf of the RPO.

4.1.4 IP commercialization and sharing the benefits

RPOs shall have procedures in place for the regular review of IP arising from their research and of the associated commercialization activities and outcomes.

RPOs shall be in a position to report to the appropriate State organisations on these activities and outcomes. RPOs and TTOs shall aim to maximize the benefits of commercialization to Ireland as a whole rather than focusing on the benefits to the RPO. They should build relationships with industry that will support a sustainable flow of commercialization outputs, rather than seeking to maximise the returns from individual negotiations.

All those involved in commercialization of IP should seek to build networks of long term knowledge sharing relationships, reflecting the ecosystem nature of innovation.

RPOs should share in the benefits of commercialisation of IP arising from their research. The commitment of researchers to commercialization and their role as entrepreneurs, taking research outcomes to the marketplace, are important and should be incentivized.

(Continued)

(Continued)

RPOs should encourage their researchers to participate in commercialization, joint R&D programmes and consultancy, through financial and non-financial incentives and rewards.

RPOs shall have arrangements in place, agreed by their governing authorities and published, for the sharing of royalties and other income from the commercialization of their IP. These arrangements should provide that income is shared between the RPO itself, the department(s) involved in the research and the individual researchers or inventors.

4.1.5 IP Management

RPOs are not in a position to provide warranties on the condition of their IP. An organisation contemplating the commercialization of IP provided by an RPO should take whatever steps it considers necessary to satisfy itself as to the condition of the IP.

However, industry is entitled to expect RPOs to have taken reasonable steps to assure that IP offered for commercialization has been managed in a professional way.

RPOs shall have policies and procedures in place that enable them, to the extent that is reasonable, to give industry an acceptable and consistent level of confidence around the management of IP arising from their research. These policies and procedures shall include arrangements for good planning, governance and execution of research programmes with particular attention to the management of publications and IP.

4.1.6 Conflicts of interest

RPOs shall have policies and procedures in place, agreed by their governing authorities and published, that minimise and manage conflicts of interest concerning the commercialization of IP and that provide guidance on doing so to their staff, contractors, consultants and students."

[Source: DJEI[1]*]*

Other issues that often arise with a new technology are whether to patent the discovery; and then whether to try to develop the technology or to license it to others for further development and use. These are complicated questions and many universities have special offices, like the Technical

Transfer Offices referred to in the document above, normally situated in the Office of Research, with experts in these areas to provide assistance.

In the United States, the Bayh-Dole act of 1980 allowed **universities to own the rights to any discovery made by any of their staff using university facilities**. However, intellectual property laws vary from country to country. This results in interesting differences between who actually holds patents in different countries. Not only are there variations in levels of ownership between firms and universities, but **in some countries individual employees can often own the IP** (as opposed to the employer organization). For example, between 1994 and 2001, in the United States, universities held 68.7% of all patents issued, whereas in Sweden during the same period, universities held only 4.9% of patents. In France and Italy the equivalent number was around 10.5%. On the other hand, companies hold 24.7% of patents in the United States but 81.1% of patents in Sweden.[2] Therefore, in order to handle the guidelines concerning intellectual property, you need to be aware of the legal situation in your own country.

In any case, for an individual researcher, the resources of the institution should be used whenever possible if questions arise about patenting or licensing any new technology.

8.4 Summary

The three topics covered in this chapter — **referencing, plagiarism and intellectual property** — are important elements in preparing a professional proposal. Using references to the current literature has many advantages in a proposal. It provides an **opportunity to contextualize your proposal** in the field and should allow you to indicate clearly why your proposal will move the field forward. It is therefore important to use up-to-date references to articles in reputable international peer reviewed journals. Judicious use of references can also save space in providing background information on both equipment and techniques.

Plagiarism, i.e., **not giving appropriate credit to other peoples work or writing**, is a major fault in a proposal and, if discovered, will lead to rejection of the proposal and maybe even involve further academic sanctions on the applicant.

Intellectual property agreements are critical in cases of collaboration between academic researchers and industry. **The sooner the agreement is reached the better**. In some cases, funding agencies will not release grant funds until an intellectual property agreement is signed. The best time to work this out is **before** any possible intellectual property is produced.

DO provide an updated list of references supporting the work proposed.

DO select references from recognized peer-reviewed international journals.

DO make certain that you provide complete and correct references so that reviewers can easily find them.

DO give proper credit to the work of other researchers.

DO prepare and have all relevant partners sign an intellectual property agreement before carrying out a collaborative research project with an industrial partner.

DO NOT reference only your own work or only local journals.

DO NOT directly copy from published work without using quotation marks and providing proper attribution and referencing.

Chapter Nine

The Budget

Don't tell me what you value, show me your budget, and I'll tell you what you value.

— Joe Biden

It's time to talk about money. Almost all of us are familiar with budgeting for our personal needs since very few of us have enough money to buy everything we would like, whether it be the latest gadget, new clothes, or travel to exotic places. We have to make choices and these choices reflect our priorities. Setting up a budget for your research project is really quite similar.

Nearly all research, particularly in science and engineering fields, requires funding for equipment, supplies, travel and to pay people. For many researchers, the greatest reward in winning a grant is that it provides the resources — the money — to allow them to carry out their research. **The budget in your proposal provides a summary of your 'best estimate' of the funding you will need** to solve the problem posed in your proposal.

The staff of the agency running a competition for funding research will determine the approximate size of a budget for each award based on the funds available. The overall funding level for the competition is normally announced in the Call for Proposals (CFP), often within a limited range, and depending on the number of awards they expect to fund. The size of the budget will provide guidance on the scope of the work expected for proposals in this particular competition. Once a preliminary decision has been made to award the grant, the actual amount of funding is sometimes negotiated with the applicants based on the budget submitted with the proposal.

While the budget can be a critical item of a proposal, reviewers may not always be able to comment on it very usefully, especially if they are not native to the country where the proposal originates and are not familiar with the salary structures, overhead agreements and the mechanisms for support of students. **Reviewers can often comment usefully on the cost of equipment and supplies and on the number of people it will take to carry out the project**.

The budget is important but in the end the award size is the responsibility of the staff of the agency. A poorly prepared budget can hurt your proposal, but generally even a well prepared budget will not help too much. Most of the experienced reviewers we interviewed expressed similar views about the budget, as illustrated in the following quotes.

> *The budget is normally not a problem unless it's so out of line that it shows that the authors do not know what they are getting into.*
>
> *I confess that as a reviewer I check the budget only for any gross problems, like overpriced items and services and missing items. The budget is, of course, a crucial element of the proposal, but it is basically the concern of the sponsoring agency.*
>
> **— Quotes from reviewers**

In a case where the award is very different from the budget requested, the applicant may need to have a discussion with the agency staff about possible modifications to the work plan.

In this chapter we will discuss different aspects of preparing and presenting a budget including, the amount of detail required, the importance of accuracy in determining the budget and the necessity to ask for what you need to carry out the proposed project.

9.1 How Much Detail Should be Provided in the Budget?

In most cases, the CFP will provide guidance about the total size of the budget and the amount of detail required, as well as the format of the budget. You should follow these guidelines carefully and provide all the information that is required. There is no need to provide more information than is asked for, particularly since reviewers will often not find this very useful. If in doubt, check with the funding agency if certain categories of cost are eligible or not. Different organizations have different policies or restrictions on the funding of travel, personnel salaries, overheads, etc.

As well as providing the total amount needed on a year by year basis for the term of the award, there is usually also the requirement to provide a breakdown into categories such as personnel, equipment, travel, supplies and overhead, as well as any other needs such as paying to use special equipment or consultants. Finally, there is almost always the requirement to **justify the expenses** in each of these categories. We will comment briefly on each of these but remember that often there is a space or word limitation applied to such budget justification so that it is important to provide the information as succinctly as possible.

In my opinion, the expenses are generally over estimated. Concerning the expenses for Personnel, there are four post doc positions and six PhD students that will perform day to day laboratory work. The objectives are clearly broad and of high interest. Nevertheless, it seems to me that the budget is excessive. The Personnel budget can be reduced. In my opinion, tasks can be reassigned so that three postdoc positions and four PhDs are enough.

The Consumables budget seems more reasonable. The global amount per year for general consumables is correct although other items, for example, analysis cost, are overestimated.

Finally, I wonder where and how many times the proposers wish to travel. The requested amount seems really excessive.

In summary, My recommendation is that the budget should be reduced to around 75% of that requested."

— Quote from a reviewer

Personnel

Generally, although it is necessary to be succinct, the more detail you can provide the better. For example, if you have definite people in mind for particular positions, give their names. This applies particularly to senior collaborators. Even for postdoctoral fellows, this information can be helpful if you already have a person on board, but who is currently being paid from another grant. If collaborators or postdoctoral fellows have special skills applicable to this grant, point this out.

You will likely also have to specify the fraction of the time spent by each person, including the principal investigator (PI), on this project, and of course this should also reflect the fraction of their salary to be paid by the award.

Graduate students are a special case. Their stipends are usually standard amounts, at least within limits, and the details of how they are paid will vary from country to country. If a graduate student is to be paid from this award, it is useful to briefly explain how much they are paid, and how much time they are expected to spend working on the grant. **Most agencies encourage support of graduate students since this is a mechanism for producing talented research personnel** to benefit the economy of the country.

Secretarial or administrative assistance is often requested in a proposal, but unless the project is particularly complex, most agencies will be reluctant to provide this support. They normally take the view that such administrative support is already present in the home organization and is taken account of, at least partly, by the overhead. However, if there are a large number of people involved in the project and it is consequently expensive and complex, most agencies will agree to at least some partial administrative assistance, possibly to be shared with the home organization.

The situation as regards the salary of the PI or faculty members on the grant can be confusing to reviewers from other countries since the rules for paying faculty will vary from country to country.

For example, in the United States, most university faculty members are paid only for the academic year (nine or ten months) and normally a grant will allow them to be paid for the summer months. In most other countries, faculty are paid an annual salary, whether they have grant funding or not, although it is possible to use grant funds to 'buy back' time from teaching to devote more time to the research project. If you think the salary requested for a faculty member or permanent staff member may be confusing to a reviewer from another country, it may be useful to add a line or two of explanation in the budget narrative.

> *Of course there can be problems if people are stupid about budgets. For example, mathematicians would sometimes put in consulting expenses for their colleagues with little or no justification, and I do not like that.*
>
> **— Quote from a reviewer**

Equipment

If you need a particular piece of equipment to carry out the project and if you do not already have access to this equipment, you will have to request funds to purchase it. Be sure to give some justification for the amount that you request from the grant. For example, one reliable source is **an actual quotation from the company supplying the equipment.** If you can have this available by the deadline for the proposal, this provides excellent justification for the amount you are requesting.

With the advent of internet search engines, it is now easy to check numbers and costs. For example, one of the people we interviewed reported that during a panel review meeting he attended, one of the proposals requested a piece of equipment whose cost seemed somewhat out of line. One of the panel members simply Googled the equipment on his laptop and found that the price requested in the budget was inflated by more than a factor of two. This was a significant black mark against that proposal.

> *Is the equipment in XXXXXX gold-plated..?!*
>
> **— Quote from a reviewer**

Another issue to consider is how the availability of equipment can impact the timescale of the project. The ordering and receiving of equipment takes time. You should show in the budget when the equipment will actually be on hand and make sure that you build this time frame into your planning (see Section 9.3).

In other cases, the equipment may be so large or expensive that its use is shared among many users. An example would be a synchrotron to provide x-rays or gamma rays for studying the structure of crystals. In this case, you will need to show that your request for time on such a facility has either been approved already, or that there is other evidence that you are very likely to be able to obtain it.

In other cases, a collaborator may be providing equipment for the project. In this case, it is useful to have some evidence, such as a letter from the collaborator, indicating his or her willingness to provide the equipment, and any constraints there may be on its use.

In these days of the universal use of computers, even proposals for theoretical work will often request computers of varying complexity in order to carry out the project. Unless there is some special advanced computer required, reviewers may question why the personal computers, currently ubiquitous in most laboratories or offices, are not sufficient for the project. **You need to take care, if you are requesting new computers, to explain why they are needed and why are they needed now**.

In some cases your project may require access to a high performance computing (HPC) facility. Depending on the policy of the owner of the HPC facilities, this may cost money which will have to be included in the budget. Even if the use of the facility is free, you should indicate your status *vis à vis* the use of the supercomputer. Have you already been allocated time? Do you have the expertise to use it?

The basic point is to provide evidence to the reviewers that you either already have access to needed equipment or that you have the means to obtain such access within the duration of the proposal.

In a climate of limited budgets, **funding agencies will often seek matching funds from the home institution or from an industrial partner to help defray the cost of expensive equipment**. In some cases, this can be a requirement for the submission of the proposal. Even if such matching is not required, reviewers will normally look favorably on the offer of matching funds since it means that the overall cost of the proposal to the agency is reduced. In the case of an industrial partner, providing matching funds for a piece of equipment indicates a serious level of commitment to the project.

> *In the first year, the PI requests an equipment budget of 134,000 Euros. While the equipment will be needed, it is unclear to this reviewer if (i) the PI is receiving a start-up package from his institution that could defray some of the costs or (ii) the PI could share the costs of some of the equipment in collaboration with other investigators who have similar needs.*
>
> **— Quote from a reviewer**

Travel

Travel is a common category in a budget and is often overused. In some cases, travel is required if it is necessary to carry out part of a project in a distant site because of equipment needs or other special circumstances. An example might be an ecological study where data needs to be gathered through fieldwork far away from your host organization. The costs of travel and *per diem* costs will need to be included in these cases.

If your collaborators are situated some distance away, regular travel for consultation may be necessary. While the use of the internet has diminished the need for some of this travel, it is reasonable to request funds for some face-to-face contact.

Travel to conferences, particularly to present results, is also a valuable use of funds. Making your work known is a responsibility of any researcher and the budget should allow for a reasonable number of presentations at scientific conferences. Conferences, especially with distinguished international speakers, also provide opportunities for sharing ideas and generating new directions for research. **However, you should be careful not to request excessive and unnecessary travel in the project budget**. Experienced reviewers can readily determine whether such travel will add substantially to the value of the project or not.

> *In the body of the proposal there is not much dis-cussion about the need for a lot of travel and then suddenly in the budget there is a major travel cost without much justification.*
>
> *The supply budget appears appropriate whereas the travel budget appears excessive (seven international trips per year for the PI). The committee may wish to consider reducing the size of the travel budget.*
>
> *Is the PI going to spend most of his time in the air?*
>
> **— Quotes from reviewers**

The reviewers thought fieldwork in the Galapagos was a bit extravagant and Darwin's travel budget got slashed.

Materials, supplies, publications

There are usually other small costs associated with a project that can be included in the budget without the need to provide excessive detail. For example, materials and supplies needed for biological experiments can be quite substantial over a period of years, and can reasonably be requested from the grant and should be realistically presented.

> *The consumables budget is underestimated for a team of that size*
>
> **— Quote from a reviewer**

Many journals require payment of the cost of publication and this is another reasonable cost to the grant, especially towards the end of the award.

Obviously, the situation will be different in different fields but as long as the costs for such items are kept within reasonable limits, experienced reviewers should not find any problems with such a budget request.

"Spare a dollar for some lab consumables, buddy?"

Overhead

Overhead (or Indirect Cost) is an amount paid to the institution by the funding agency on all grants obtained by the staff of the institution. The direct costs are items explicitly listed in the budget such as salaries, equipment, materials and supplies, and travel. The overhead is typically a fixed percentage of the direct costs, and is used by the institution to cover the cost of providing infrastructure (buildings, utilities, maintenance, cleaning) plus the costs associated with hiring and retaining research personnel. The amount of the overhead is usually based on an agreement between the institution and the funding agency, and the negotiated rate applies to that institution for a number of years. The overhead cost is normally a percentage of the total amount of the grant (sometimes minus the equipment purchase costs). and is subtracted by the institution before the grant funds are made available to the PI. The overhead rate agreement is usually stated in the CFP document and is normally included in the total award for the project.

In the USA, the Department of Health and Human Services often negotiates the overhead rate on behalf of other agencies.[1] Most other countries have similar procedures to assist institutions with the cost of supporting research by their staff.

In the United Kingdom, many funding agencies require applicants to indicate the entire economic cost of a proposed project — an exercise called 'full economic costing' (FEC).[2] The FEC total is supposed to include all possible research expenses including the costs of the time of participating tenured staff, the use of the host organization's facilities and indirect costs. UK research funders normally then pay a percentage of these costs as part of the grant. The UK Research Councils typically pay 80%, but some charitable research foundations pay significantly less. The FEC approach is intended to help ensure university-based research activities are carried out on a sustainable basis, with appropriate ongoing investment in the upkeep and maintenance of buildings and facilities, and proper accounting for staff time.

Including the overhead in the total award can sometimes lead to disgruntled researchers who resent the fact that it appears that they are required to pay overhead from 'their' grant funds. However, they are often not aware of the very real costs imposed on the institution from the extra work and facilities required to carry out research. In fact, overhead is intended for use by the host institution and not the researcher. It is normally awarded as a separate agreement between the funding agency and the research institution, to compensate the institution for the cost of hosting the project.

Perhaps the resentment also springs from the fact that some institutions have, historically, returned a fraction of the overhead to the department or even the individual PI who generated the award, as an incentive to encourage research activity. With many universities becoming ever more research intensive and with increasingly tight university budgets, the practice of returning part of the overhead is rapidly disappearing.

Of course, in some cases the overhead costs can be excessive and there is a continual discussion and regular negotiation of the overhead rates between the agency providing the funding and the institutions at which the research is carried out.

While this may seem complicated to the individual researcher submitting a proposal, **the research office of your institution will be very familiar with the rules regarding overhead or indirect costs and will be able to provide assistance in calculating the amount** to add to the direct budget if this is required. It is always a good idea to seek help from research officers at your university if you have any questions about setting up the budget.

9.2 Accuracy

Even though a budget is simply a document that estimates the total cost of a project, usually broken down by category and time, most agencies will use the budget numbers compared to actual expenditures in checking on the progress of the project. Therefore, it is important to provide budget numbers that are reasonably accurate and that take account of the progress of the project over time.

Reviewers will also be looking at the budget numbers — including the number of people involved and the cost of equipment and travel — to make sure that the costs in any of these categories are not excessive. **If they believe that the project could be carried out with significantly less funding, they will point this out and generally will score the project more negatively as a result**.

Another obvious way in which you can give a negative impression to the reviewers is to have arithmetic errors in the budget. There are many possible sources of error, from simple calculation mistakes to wrong assumptions. Whatever the cause, errors that creep into the budget as presented in the proposal will be interpreted by the reviewers as carelessness on the part of the PI. The solution is to check and double check the numbers in the budget and, if possible, have the numbers checked independently by another person.

9.3 Year by Year Budgeting

In most cases, where awards extend for more than one year, the applicant is asked to state the anticipated rate of spending year by year. One simple approach is to allocate the total funds equally each year, but in reality this would be a very unusual spending pattern. For most projects, equipment needs to be ordered and people have to be hired in the first year. These things take time. People often need time to relocate and it can take more than one year for equipment to be delivered and installed. Therefore the first year costs are normally less than ongoing costs.

Similarly, travel to meetings to present results is more likely to happen later in the project, so travel costs may peak towards the end of the award.

Each project is different, but you should give some thought to the time development of the budget. A well thought through budget which does not simply divide the total cost by the number of years, will reflect well on the applicant and will be appreciated by the reviewers. A rational spending plan will also be helpful to you when the funding agency asks for an update of your spending during the course of the award.

9.4 Should you Ask for More Than you Really Need?

The very short answer is NO!

There is sometimes a mistaken belief that there is an advantage in 'padding' a budget by asking for more funding than you need to carry out the project. The argument is that the funding agency will typically reduce funding for all projects by a fixed percentage so that asking for more initially means you will have enough resources to carry out the project even after a reduction.

However, only the most unsophisticated funding agencies actually reduce funding for all proposals based on a formula. Rather, using the expertise of the agency staff and the comments from reviewers, funding agencies try to provide sufficient funds to allow the project to be successful without providing any surplus funding which could be better spent else-where. If the project is seen as unjustifiably expensive by the reviewers, they will take this to reflect on the expertise of the PI and will generally score the project lower. In some cases, reviewers may even see this as a 'fatal flaw' in a proposal, which would then lead them to recommend to the funding agency that it should be declined.

The best strategy is to ask for what you need to carry out the project successfully, but not more. Try to make the best estimates of costs, and present your justification or rationale for the numbers whenever possible. This transparency will generally be well received by reviewers.

> *The program could be done by the PI + 1 student ... the army of post docs requested is excessive...*
>
> *If a young person submits a budget that is very large, given their limited experience, this will not go over well. It just shows a lack of realism.*
>
> **— Quotes from reviewers**

9.5 Example of a Budget

The following is an example of a budget taken from an actual proposal. This budget, for a three-year project, illustrates some of the points mentioned in the sections above. These include:

- The staff budget mentions specific people when these are on board.
- The staff budget varies from year to year as people are recruited or leave.
- The equipment budget is front loaded.
- The materials budget shows some ramp up and then a steady level of expenditure.
- The travel budget is linked to travel by individuals on the staff and is the smallest item in the budget, behind equipment and materials.
- Publication costs peak towards the end of the project.

Description	Year 1	Year 2	Year 3	Total
STAFF	459,477	518,321	419,219	1,397,017
EQUIPMENT	53,000	4,000	10,000	67,000
MATERIALS	37,000	50,000	53,000	140,000
TRAVEL	13,500	15,000	13,500	42,000
TOTAL DIR COST	**562,977**	**587,321**	**495,719**	**1,646,017**
STAFF				
Liam A (PI)	135,715	127,233	84,822	347,770
Sean B (Res Fellow)	79,758	86,209	92,659	258,626
Eoin C (Res Fellow)	79,758	86,209	92,659	258,626
Gary D (Sen Post Doc)	69,701	71,611	0	141,312
Betty Li (PhD + postdoc)	25,000	50,493	0	75,493
PhD student 2	25,000	25,000	25,000	75,000
PhD student 3	0	25,000	25,000	50,000
Research Assistant	40,545	42,566	44,586	127,697
part of Research Assistant	4,000	4,000	4,000	12,000
Total	**459,477**	**518,321**	**368,726**	**1,346,524**
EQUIPMENT				
Nanodrop	9,000	0	0	9,000
Ultra centrifuge	20,000	0	0	20,000
PCR hood	5,000	0	0	5,000
Orbital rocker	2,000	0	0	2,000
2xsets Gilsons	4,000	0	0	4,000
Computers	4,000	2,000	2,000	8,000
Benchtop Centrifuge	0	0	3,000	3,000
2xVartex	2,000	0	0	2,000
2x PCR machine	5,000	0	5,000	10,000
2x 20 freezers	0	2,000	0	2,000
3 x fridges	2,000	0	0	2,000
Total	**53,000**	**4,000**	**10,000**	**67,000**

(*Continued*)

(Continued)

Description	Year 1	Year 2	Year 3	Total
MATERIALS				
Sequencing	10,000	14,000	14,000	38,000
Oligonucleotides	8,000	12,000	12,000	32,000
Gene synthesis	3,000	3,000	3,000	9,000
Restriction Enzymes	1,000	1,000	1,000	3,000
Antibodies	2,000	2,000	2,000	6,000
Mass Spec Analysis	5,000	5,000	5,000	15,000
Publication costs	3,000	7,000	10,000	20,000
Protein purification kits	5,000	6,000	6,000	17,000
Total	**37,000**	**50,000**	**53,000**	**140,000**
TRAVEL				
Liam A	4,500	4,500	4,500	13,500
Sean B	3,000	3,000	3,000	9,000
Eoin C	3,000	3,000	3,000	9,000
Gary D	1,500	1,500	0	3,000
New Postdoc	0	0	1,500	1,500
Betty Li	1,500	1,500	0	3,000
PhD student 2	0	1,500	0	1,500
PhD student 3	0	0	1,500	1,500
Total	**13,500**	**15,000**	**13,500**	**42,000**

9.6 Summary

The budget is an important part of the proposal and you should take care to develop one that will provide you the resources to carry out the work as proposed. The CFP will normally provide a definite format and space limitation for the budget. Make sure that you follow the guidelines precisely. You will usually have to provide succinct but clear justification for the amounts requested in each category. In cases where reviewers are not local to the country, they may not be able to comment on some parts of the budget, such as the cost of personnel. However, they will usually be familiar with the cost of supplies and equipment and on the amount of travel requested in the budget. If your project calls for using a large and/ or expensive piece of equipment not under your control, you will need to explain how you plan to obtain access to this equipment, either through a separate time request to a user committee of the facility in question or through a collaborative arrangement.

- **DO** make sure to use the format required in the CFP for the budget and keep within any space constraints required.
- **DO** provide justification for all items in the budget as far as possible, especially large equipment items.
- **DO** give relevant information on the expertise of personnel if this is available when the proposal is submitted.
- **DO** think about the time frame of the budget and how much you will actually spend each year.
- **DO** check the budget and recheck it for accuracy. Arithmetic errors will be construed as carelessness by the reviewers.
- **DO NOT** request unnecessary travel.
- **DO NOT** 'pad' the budget. Ask for what you need, but not more than you need.

Chapter Ten

Addressing Reviewers Comments

Is there anyone so wise as to learn from the experience of others?

— Voltaire

Oops! You have just learned that the reviewers didn't love your proposal as much as you expected. Your first reaction, very naturally, may be anger and disappointment that all your work in writing the proposal has been in vain. **Don't those reviewers appreciate a good idea when they see it?** After you have calmed down, perhaps in a day or two (or three), you will realize that there is almost always something useful to learned from the reviews.

The almost universal practice of research funding agencies is that when the review process is completed, the anonymous reviews of each of the proposals is shared with the authors of the proposals. These reviews may be redacted slightly to preserve the anonymity of the reviewer, or other colleagues or institutions mentioned in the original reviews. In most cases, this is the end of the process and no rebuttal of the reviews is possible.

However, there are two different situations where you may have the opportunity to respond to reviewers comments.

1. The agency may provide you the opportunity to respond to individual reviewers comments before further consideration of your proposal either by the Program Officer or more commonly by a separate panel

of reviewers. Your response is not required and normally has to be submitted within a very short time.

2. Your proposal is declined and the comments of the reviewers are sent to you for your information. If you wish to submit a similar proposal to a later competition it is in your interest to carefully consider the comments of the reviewers.

10.1 Responding to Reviewers Within the Peer Review Process

Some research foundations will give applicants the opportunity to respond to postal reviewers' comments. Typically, responses will be invited in advance of a panel review meeting, where the panel considers all of the 'evidence' — proposal, reviews and response — before making their recommendation for funding.

Applicants are normally given the chance to answer any direct questions raised by the reviewers and identify any statements by reviewers which they believe are factually inaccurate. Sometimes, if reviewers have taken the time to suggest areas where the project might be improved, there may be an opportunity for you to say whether you are going to take the recommendations on board. **Applicants are typically not given the opportunity to debate the value judgments of the reviewers**. This is generally unwise in any case.

Giving applicants an opportunity to respond to reviewers comments is time-consuming and labor-intensive, and many funding agencies view it as inefficient and impractical. Nevertheless, with enough time and resources, it can be a useful part of the peer review process, giving more information to panel members about issues they might have raised themselves in any case. A high quality response can make all the difference between a borderline proposal being funded or not.

There is normally no obligation for an applicant to respond to reviewers' comments. If given the opportunity, however, it is almost always a bad idea not to respond, especially if the reviewers have raised very clear questions designed to reveal whether the proposal is fundable or not.

How **_not_** to respond to reviewers.

10.2 How to Respond

Stay calm. Responding to reviewers in an angry or hostile way almost always comes across as defensive, arrogant or poorly considered. Don't forget that the agency program managers and the review panel will know who the reviewers are and that they are likely to be very distinguished in their field. Dismissing the reviewers' comments as being simply wrong or questioning their credentials can make you look very foolish if their track record and credentials are impeccable.

Things that may help convince reviewers

- **DO** use a concise and measured tone.
- **DO** answer ALL questions raised by reviewers.
- **DO** structure your response in a systematic way using clear headings.

- Where appropriate, **DO** use references to peer reviewed publications to correct any factual inaccuracies.
- **DO NOT** take an argumentative tone.
- **DO NOT** question the credentials of the reviewers or dismiss their opinions.
- **DO NOT** argue that the strengths of other aspects of the proposal (e.g., potential impact) outweigh some of the weaknesses identified by the reviewers.

10.3 Reviewers Comments Following an Application Being Declined

When you receive a letter from the research agency telling you that your proposal has been declined, you will naturally be disappointed and perhaps even angry. As with most tense situations, you are probably best advised **not to respond in any way too quickly** but instead put the reviews away for a day or two before looking at them more carefully.

The first thing to realize is that having a proposal declined puts you in very good company. Almost all scientists, including Nobel Prize winners, have had proposals declined. Stephen Chu, Nobel Laureate and US Secretary of Energy, told a large audience at an American Association for the Advancement of Science meeting in 2014 that he had had two proposals declined within the past year. This happens to everyone.

Next, read over the reviewers' comments carefully. These are fellow scientists who have taken the time and made the effort to read your proposal and have provided their best judgment on its merits. Therefore you should take their comments seriously. Use this as a learning experience and think carefully about the views expressed by the reviewers. You may not agree with all their opinions, and there may even be contradictory ones expressed by different reviewers, but still you need to consider them.

One reviewer comment that can be particularly galling is a statement that 'the proposal is not clear'. You may believe that your proposal is perfectly 'clear' and that the reviewer has just not read it carefully enough. However, you can be sure that the reviewer has, in almost all cases, made a genuine effort to understand your proposal and the fault is with you. So read over your proposal again or, better still, have a critical friend read it over for clarity. Normally you will find that you can clarify either the motivation or the methodology so that your purpose and your approach are presented more clearly.

Another common statement from reviewers, especially in the case of less experienced applicants, is that the proposal is too ambitious and tries to do too much with the resources available. This is a very useful observation. You may choose to respond by dialing back on your aims and choosing to resubmit a more modest proposal, realizing that the project may take more time than you anticipated. Nevertheless, ambition is laudable, so don't cut back too far or your proposal may become mundane, and may not push the boundaries of knowledge enough.

Alternatively, you may choose to keep the same goals but demonstrate much more clearly and explicitly in a new proposal that you have the ability to carry out the goals in the time and with the resources available. This may mean obtaining some preliminary results or pointing out that in the past you have succeeded with projects of similar scope. **Reviewers need to be convinced that you can accomplish the goals that you propose**.

Usually, it is also valuable to contact the Program Officer to discuss the reviews with him or her. In the case of a summary from a panel, the Program Officer will have been present at the panel discussion and may be able to give you more insights into the meaning of some of the comments in the reviews. Listen carefully to any clarifying comments from the Program Officer since these will give you some indication of his or her priorities. Do not be argumentative or defensive if possible. Remember that you are **seeking information to help you prepare a better proposal** for the next competition.

10.4 Resubmission

Having read over and pondered the reviewers' comments, you should be in a position to revise your proposal for a future and hopefully improved resubmission. A word of caution here — do not submit the same proposal without any changes simply because you think that your original proposal was fine. Some funding agencies do not allow you to submit exactly the same proposal more than once. At a minimum, you will need to update the list of references. And you would be very wise to take account of the reviews to prepare a stronger proposal. You do not need to agree with all the comments but you need to take them into account in making a stronger case when revising your proposal.

Remember that you are receiving a free critique of your proposal from experienced people in your research area. At the very least these are people who are trying to help you write a stronger proposal and often will have some excellent suggestions for how you can improve your plan or your presentation. In some cases they will make suggestions of extensions or additions to your work that you may not have considered. So do take their suggestions and comments seriously.

Your resubmitted proposal may be sent to some of the same reviewers who reviewed your original proposal. There may also be some new reviewers. Therefore you can probably assume that there will be someone reading your proposal who will be aware of your responses (or the lack of them) to previous criticism.

Finally, it can be very unsettling when criticism of your new proposal contradicts comments made by some of the original reviewers. Just take this as a sign that reviewers are fallible and do not let it discourage you. Generally, there is a good deal of consistency among reviewers and you just need to be persistent. The old saying that "If at first you don't succeed, try, try again" is very applicable to grant-writing.

DO take some time to allow your disappointment to subside before reading the reviewers comments carefully.

DO treat the reviewers' comments as a means of improving your next proposal.

DO take the opportunity to respond to reviewers' comments if offered it.

DO contact the Program Officer to learn as much as you can about the review of your proposal.

DO NOT submit the same proposal without updating the content and the references.

Chapter Eleven

Special Grant Competitions

There are two kinds of people, those who do the work and those who take the credit. Try to be in the first group; there is less competition there.

— **Indira Gandhi**

Almost all of the discussion in the earlier chapters has been about a generic research grant proposal, either for an individual award or an award to a group of researchers. There are, however, other sources of funding that can move your research forward in different ways.

There are many different kinds of awards available which may be appropriate in particular cases. **Therefore you need to be on the lookout for awards which might be tailored to a particular project that you have in mind** (or that might enhance an ongoing project), even if they are not discussed here.

In this chapter, we outline some of the most useful categories of special awards that researchers should be aware of. Many of the principles that we have pointed out in earlier chapters also apply to these special grants, although there are extra aspects you will need to consider in certain cases.

There are often situations where a funding agency wishes to encourage a particular activity, for example international collaboration, or is convinced that there is a particular need in the local research community that is not being met, such as enough modern instrumentation. In such cases, the agency will arrange a competition focused on that activity or issue. We will consider a number of these special awards designed to meet particular needs, including grants:

- For instrumentation.
- For travel, particularly international travel.

- To promote international cooperation.
- To assist early career investigators.
- For education and outreach.
- For hosting workshops or summer schools.
- To promote the development of research resulting in commercial undertakings, or generally.
- To encourage engagement with industry and knowledge transfer.
- To help researchers return to a research career (following an absence due to family or other caring responsibilities, long term illness, etc.)

A particular class of special awards is large 'center' grants involving multiple principal investigators (PIs) and groups, often from different fields and sometimes from different institutions. These were discussed in detail in Chapter Six.

In some cases, special awards are used to promote more diversity in a specific area of research, or to encourage people in underrepresented groups to consider research careers. Typically, these competitions are similar to regular competitions, but the applicant pool is limited by gender or sometimes by ethnic group. Since the rules for such competitions are similar to those for regular grant competitions, we will not consider them further here.

The activities promoted by these awards can, in some cases, be funded from regular research awards. For example, foreign travel, education, outreach or knowledge transfer can in principle be paid for by grant funds. However, in many cases, the funding available in a research grant does not allow sufficient attention to be paid to some of the special requirements that the agency believes are important. As we will see in Section 11.1, large scale instrumentation used by a number of groups falls into this category.

11.1 Instrumentation Grants

While normal project grants often request special equipment in the budget, the funding for such equipment is often limited, meaning it may not be possible to purchase very large items by this mechanism. Often large pieces of equipment, such as an electron microscope, are shared between different groups and are used by many different projects. Therefore, a funding agency will sometimes run a special competition only for instrumentation, and especially for large pieces of equipment that cannot easily be accommodated in a normal project budget. Such funding provides an opportunity for research institutions to improve their scientific equipment and make it more up to date, allowing researchers to undertake more ambitious projects. As a result, equipment grant competitions are usually very popular with both research institutions and individual researchers.

Advice to funding agencies

There is also a further advantage of such awards to the agency in managing its yearly budget. Normally project grants run for a number of years so that these awards commit funds from the agency for a number of years into the future. However, **an equipment grant is a one time award** and normally will not commit the agency's funds beyond a year. This can be an advantage in using up any residual funds remaining in the agency's budget at the end of the fiscal year without generating any longer term multi-year commitments.

There may be a temptation for a research institution or even individual researchers to take the attitude about equipment grants that 'if we build it they will come!' However, equipment grants are typically reviewed like regular project grants and so the **need for specific equipment must be articulated clearly and the new science that can be accomplished must be spelled out**. The best way to do this is by briefly outlining a number of projects that could be carried out with the new equipment and arguing that they could not be carried out without it, either because of the excessive time required to acquire the needed data, or because of the limited accuracy of the currently available equipment.

In a number of cases where large pieces of equipment are sought, there will be **multiple users**, sometimes from the same department or the same institution, or sometimes from a number of different institutions. This is normally an advantage in seeking instrumentation funding since the agency will want to have expensive equipment used by as many researchers as possible.

However, in such cases, the management structure of the equipment facility will need to be laid out in the proposal, along with information on how time will be allocated and criteria for such allocation. Will there be a charge for using the equipment? How is the charge determined? Will training be required before researchers are permitted to use the equipment or will the costs include operators to assist researchers? If operators are to be provided, who will pay for them? These are all questions that need to be addressed in the proposal to indicate that the equipment will be used efficiently.

Normally overhead is not charged on equipment, whether as part of a project grant or a special equipment grant (although this can happen if using the equipment incurs exceptional costs, e.g., using extremely large amounts of electricity). However, for large pieces of equipment requested in a special equipment grant proposal, there is often a requirement for **matching funds**, sometimes as large as 50% of the total award. Matching may come from the research institution as a contribution to infrastructure improvement, or it could also come from an industry partner.

In a proposal for an equipment grant, it is important to establish the cost of the equipment as accurately as possible. Reviewers will be alert to any attempt to inflate the budget request for equipment. **Perhaps the best way to determine the cost is to obtain a quote or a number of quotes** from different vendors, which then establishes the cost without any ambiguity. In some cases, specialized equipment can only be purchased from a single vendor, but again a quote from the vendor that is firm for a certain time will be helpful to give credence to the budget. In some cases, vendors of scientific equipment are willing to give discounts to educational institutions, since it is to their advantage to have their equipment used by young researchers at the institution. This possibility is always worth discussing with a vendor.

Finally, the cost of operating the equipment **including regular maintenance** needs to be in the proposal. The funding agency is unlikely to be willing to fund the operation and maintenance of the equipment, at least beyond the first year or so after purchase. Therefore, it will be necessary to set up a business plan to cover the cost of maintenance and operations. Possible sources of ongoing funding might be, for instance, the research office of the institution housing the equipment, or from user fees. It is critical, however, to make this clear to your host institution and get agreement in advance. You do not want to take on responsibility for equipment which will be a long term drain on the resources of your research team or your host institution.

DO identify new research activities that can be carried out with the new equipment being requested.

DO expand the group of possible users of the new equipment provided this is appropriate and realistic.

DO try to obtain matching funds for the equipment requested either from your home institution or industrial partners.

DO NOT neglect the need to provide a management plan for the use of new equipment, possibly including operation and maintenance.

DO NOT inflate the price of the equipment proposed. Be realistic and obtain quotes if at all possible and present these with your proposal.

11.2 Travel grants

Most project grants provide some funding for travel, often to present the results of the research or to engage with collaborators. However, in some cases agencies are willing to provide special travel grants for specific purposes. We will discuss a few examples of different kinds of travel grants and say a little about the review of these types of grants.

Conference travel grants

In some cases, an agency might be willing to support travel to large international conferences even if the researcher does not currently have a project grant. The agency might wish to support the conference, or it might believe that there are benefits in having local researchers present their work at an international conference. Or the agency might wish to stimulate interactions between local researchers and the international community. These grants are often quite small but can have large payoffs in generating new research ideas and stimulating research activity locally.

Such awards may not pay the full cost of the visit so it may need to be matched by the local research institution. Usually travel grants are reviewed internally by agency staff since they normally do not involve large amounts of money.

A researcher is far more likely to be successful in such a travel award competition if they are presenting a paper at the conference. This means planning ahead to have a paper prepared and to make certain that you meet any conference deadlines for submission.

Grants to allow researchers to spend time in a foreign laboratory

Another type of travel grant which is not normally covered by a project grant is one that allows researchers to travel to a foreign country and spend time in one or more different laboratories. The purpose of such visits can be to learn a new technique or to gain experience using new equipment. Or it can simply be to gain exposure to new ways of thinking about issues in the field. Such visits, for times ranging from three months to a year, can also result in useful research collaborations and expand the scope of the equipment available to local researchers.

Comment from one of our interviewees:

In one proposal that he reviewed, he noted from the budget that the principal investigator (PI) and a postdoctoral fellow were scheduled to spend some time in a laboratory in Japan. This made sense to him because this is where people from less developed science infrastructure countries can learn forefront techniques and bring these back home. It also gives them exposure to ideas. He would suggest that this is useful in many other proposals.

Since these awards include both travel costs and a stipend for the time spent in the foreign country, they will be more expensive to fund than a simple travel grant to a conference. Therefore, they are more likely to be reviewed external to the agency.

When applying for such awards it is important to explain carefully the **benefit of the award not only to you, the individual researcher, but also to the local research community**. For example, when you return, you may be willing to offer a short course on new techniques that you have learned. You should certainly be willing to give a seminar outlining what you have learned and how it might be applicable locally. A research funding agency is always seeking ways to expand the research capabilities of the whole research community which it serves.

Grants to allow foreign visitors to stay for extended time

A reverse kind of award to the one described above is where grant funding is made available to bring distinguished foreign visitors to the country for an extended time, ranging from perhaps one month to one year. This kind of award can be stimulating to the researchers in the institution where the visitor spends most of his or her time. Such awards are sometimes used as a recruiting tool to attract foreign scientists. Having a distinguished foreign visitor spend a significant amount of time in the country also serves as a showcase for the local research, and is one method of improving the international awareness of national activities and strengths as well as promoting the connectedness of the country. Depending on the size of the country and the distribution of scientific expertise, there may be advantages in encouraging a visitor to spend some of his or her time at more than one institution.

As mentioned earlier, in some cases where the amount of a travel grant is small, the application process is simplified and the review carried out within the funding agency, without the involvement of external peer reviewers. However, when the awards become larger — whether because a local researcher spends a significant period of time overseas or a foreign-based researcher spends longer in the host country — there may be a more complex application process required, often involving peer review.

The typical criteria used in evaluating these awards is twofold — why the award will be of value in improving the research in the home country, plus the track record of the scientist obtaining the award. **Therefore, it is important to explain clearly what anticipated benefits are likely to accrue because of the visit**. In particular, it may be worth highlighting the complementary or synergistic nature of your research with that of your international partner. If there is already a history of external collaboration resulting in publications, this will be an excellent positive indicator for future awards.

In seeking funding to attend a conference, **DO** make sure that you present a paper or at least a poster at the conference.

In seeking funding to spend time in an overseas laboratory, **DO** explain clearly the benefit of such a stay, not only to yourself, but in addition to the local research community.

In seeking funding for an international visitor, **DO** encourage the visitor to spend some time in different institutions if possible.

11.3 International Cooperative Agreements

Beyond the case of a single researcher spending time in a foreign laboratory, there are situations where groups of researchers in different countries collaborate on a research project. This is fairly common, and indeed the US National Science Foundation has a whole division in the organization devoted to international activities, including collaborative agreements. The usual procedure is for each group to submit the same proposal to the appropriate funding agency in their own country. These proposals should lay out the role played by the different groups, including the budget required from each agency, and explain why the collaboration adds value to the project.

Again, review criteria will be similar to conventional research grants, but with an additional focus on the added value of the collaborative partnership. Sometimes international research collaboration initiatives are based around particular strategic challenges where the countries involved have complementary strengths. Such strategic programs may have additional criteria relevant to the particular challenge being addressed.

As discussed in Chapter Six, there are many reasons why such international collaborative programs are undertaken. In some cases, one group has specialized equipment that the other group would like to use to carry out the project. If both groups have an interest in the project then they may seek funding from their respective funding agencies under an international cooperative agreement. Alternatively, it may be that one group has, for example, access to special data or samples and the other group has expertise in various kinds of analysis, so that a collaboration between the groups would be more likely to achieve results. There are many, many examples of such international collaboration. One example, for illustration, is the case of the Iceman.

In 1991, the mummified corpse of the 'Tyrolean Iceman' was found frozen in a glacier near the Austrian–Italian border. Currently the body is located in a special museum in Bolzano in northern Italy. The Iceman was determined to have lived about 5,300 years ago. The complete mitochondrial genome sequence of the Iceman was established by a collaboration between two Italian institutions, the University of Camerino and the Consiglio Nazionale delle Ricerche, Milano, as well as the University of Leeds, England.[1] All three institutions brought expertise to the project that made for a better outcome. Many similar examples of international collaboration exist in a variety of scientific fields.

One of the many difficulties in such agreements is the different review processes in different countries. The normal pattern is for each country to review the proposal using its own guidelines. In this case, all the agencies must reach a conclusion to fund before the project is funded by any one agency. However, it is possible that the funding agency in one country may decide that the peer review of the project in another country is sufficiently rigorous that they will waive their own review and abide by the conclusions of the review in the other country. Of course this funding agency would still have to agree to contribute the requested budget to the researchers in their home country to make the project viable.

Advice to Funding Agencies

The setting up of such international agreements can be complicated since the programs often require agreement between various government agencies in more than one country. For example, in the US–Ireland agreement, that was initially restricted to only a few areas of science, there were at least six agencies involved in the discussions, which took about two years to complete. However, international agreements are seen as sufficiently worthwhile that many countries actively participate in them.

11.4 Early Career Awards

There is often the belief that young researchers are at a disadvantage compared to their more experienced colleagues in competing for grant funding. They do not have as long a track record and their experience is necessarily limited. Therefore some funding agencies have introduced special competitions which are restricted to 'young' or more typically 'early career' investigators. ('Early career' is occasionally defined by age, but more commonly by the number of years beyond a PhD– normally around 3–7 years). The term 'early career' is generally preferred to 'young', as such awards typically also allow applicants who are older but, for example, may have spent time in industry before returning to academia. Sometimes these competitions are run in combination with a general competition, with a certain number of grants or a certain amount of money designated for 'early career' investigators. These awards are typically fairly modest, given that the early career researcher will not normally have experience in managing a large team.

Another use of 'early career' awards is as a recruiting tool. In many countries without a consistent history of research funding, many young scientists have left their homeland to seek careers in a more established (and better funded) scientific culture. However as countries are trying to build their scientific establishment, one source of new active scientists is their own citizens who have begun their careers elsewhere. The funding agency and the research institutions within a country will therefore try to work together to attract these young people to return to their home country to continue their careers. One such mechanism is these young investigator award, since it is easier to attract someone who has only recently left and who is still establishing her or his career, rather than a more experienced researcher who may have a more established life and career in another country.

The awards in these cases are very prestigious, often bearing titles like 'Presidential Young Investigator, and carry significant research funds.[2] They are also quite competitive and reviewed extensively, often requiring a visit from the applicant and a presentation before a panel of reviewers. Such awards may not be restricted to citizens of the home country since recruiting young scientists from foreign countries can also be an additional goal of these programs. However, the ex-pat community is usually the most fruitful source of applicants.

Because these 'young investigator' awards serve as 'development' or 'starter' awards for beginning researchers, the agency will often look for a broader program with longer term goals than would be expected in a normal single investigator grant. In the US, it is often the case that an educational component is also sought.

An example of a special Career Development Award is given in the following block, showing the preamble for the 2013 Award program at Science Foundation Ireland (SFI). This is an illustration of special four-year awards restricted to beginning researchers to help them with their early careers. This preamble is taken from the SFI Web Site.[2]

SFI CAREER DEVELOPMENT AWARD PROGRAMME 2013

SFI is committed to supporting and developing early- and mid-career researchers with the greatest potential to become excellent, fully independent research leaders and offers a suite of funding opportunities to help make this transition. The **new SFI Career Development Award (CDA) Programme** supports excellent investigators still in the earlier stages of their research career who are already in an independent (either permanent or fixed-term) academic position. The award has a four-year duration and is intended to provide award holders with the opportunity to extend their research activities by allowing research teams to be built or expanded, and to assist in the procurement of required items of equipment and consumable materials to carry out the planned activities.

Based on our experience in dealing with these 'Young Investigator' Awards, there sometimes appears to be a temptation to exaggerate one's achievements. This should be resolutely resisted. Reviewers will react very negatively to any perceived exaggeration. Reviewers are aware that the candidate is at an early stage of her or his career and will assess their achievements accordingly. It is unnecessary to embellish what are often already remarkable successes.

"And then I won the gold medal of the Society of Advanced Carpentry and Artificial Intelligence...."

In addition to all the issues already discussed in preparing a normal research proposal, we would offer the following additional suggestions:

DO be scrupulously accurate in presenting your achievements.

DO NOT attempt to pad your CV.

DO take care to practise any oral presentation that may be required, including taking account of the time allocated to any presentation.

11.5 Education and Outreach

Most government research funding agencies are interested in increasing the number of scientists and engineers in the country. Therefore, they wish to encourage students to choose careers in Science, Technology, Engineering and Mathematics — the so-called STEM areas. But in order to graduate students in STEM subjects with advanced degrees, there needs to be significant numbers of students interested in these areas in high school and in undergraduate programs. There are a number of special programs to address this issue. In addition, many research grants, especially large center grants, are strongly encouraged to engage in outreach to the community, especially to high school students or even students in elementary schools.

In the USA, one of the most popular and successful of these programs is the Research Experience for Undergraduate (REU) program administered by the National Science Foundation (NSF). The REU program provides the opportunity for undergraduate students to participate in research programs in both university laboratories and in national laboratories, mainly during the summer months. Normally there are also experiences common to all the students such as advanced classes or field trips. There are normally a number of faculty involved in the program, each of whom will supervise one or more undergraduate students. In 2014, there were around 660 active REU programs in a variety of STEM disciplines across the US.

An application for an REU grant has a number of special elements. As well as requiring strong research programs being carried out by the participating faculty members, **the proposal must describe the undergraduate mentoring experience of the faculty**. These programs have become so popular with both students and university departments that participation in an REU program has become almost a 'right of passage' for an undergraduate wishing to pursue a PhD degree in a STEM area.

Be aware, however, that the primary purpose of an REU grant is to provide the student with a valuable learning experience. Although undergraduate researchers can make significant contributions to research projects, the grant is intended for education and training purposes and not really to provide you with additional research support. In fact, if anything, supervising undergraduate research projects can be time-consuming and a distraction from your core research activity, so make sure you take on REU students for the right reasons.

In Europe and Asia there are also many programs for undergraduate students to participate in research. Many of these are supported privately and most require direct student application to the department or institute. For example, the European Molecular Biology Laboratories (EMBL) offers undergraduate students in many disciplines the opportunity to carry out research during the summer at one of their sites in Heidelberg, Grenoble, Hamburg, Hinxton or Monterotondo.

11.6 Workshops and Networking Grants

Sometimes a branch of a science field, or more often a new linkage between fields, is ripe for rapid advance. A funding agency will often be willing to fund a workshop around such a topic to bring together the important players and to help move the area rapidly. **Researchers should be aware of this possibility and be prepared to make the case for a workshop when such opportunities arise**. There are many historical examples of such cross-disciplinary activities that have been responsible of breakthroughs in understanding. Perhaps the most common one recently is the link between physical and biological sciences leading to new analytical methods. Or the link between mathematics and genetics leading to advances in understanding the causes of certain diseases.

Other types of workshop include 'summer schools' or similar events designed to support graduate student, and post doc training events in emerging research domains. Applicants for this type of workshop must make the case for the strategic value of such training events and the current gap within the research community.

Typically, applying for money to supplement established workshops or conferences with existing funding streams will not be as compelling to the funders or to the reviewers.

Workshops with participative sessions, e.g., breakout working-groups and interactive panel sessions, are generally better regarded by reviewers than 'talking shops'.

DO plan to generate outputs in the form of workshop reports (or other materials) to ensure the insights and findings of the event can be disseminated more broadly to the community.

DO involve early career researchers (including PhDs and postdocs).

DO give names of the workshop organizing/advisory committees.

DO give details of plans to publicize the meeting.

11.7 Industry Engagement and Knowledge Transfer Awards

One commonly stated purpose of fostering scientific research in a country is the hope that a vibrant scientific community will create useful commercial products from research discoveries. There is ample evidence of this happening in many cases where small companies,[a] which later became giant ones, have grown out of laboratory discoveries. Hewlett Packard is just one case. In order to encourage this transfer of knowledge into the marketplace, funding agencies often set up specific awards to promote such transfers.

One of the most successful of these, called the Small Business Innovative Research (SBIR) program, began in the late 1970s at the US National Science Foundation (NSF). This program and similar programs like the Small Business Technology Transfer (STTR) program have expanded to a number of US Federal granting agencies. A useful summary by the Small Business Administration can be found at their web site.[3]

The main purpose of these programs is to stimulate technological innovation and to increase private sector commercialization. The SBIR program is designed to allow small businesses access to Federal research and development (R&D) funding in cases where there is potential for commercialization. The STTR program is designed to facilitate cooperative R&D between small businesses and US research institutions, again where there is potential for commercialization and may be of more interest to university faculty members.

The criteria for review of SBIR/STTR type programs involve not only scientific merit and the track record of the participants, but also some measure of the likelihood of commercial success.

Advice to funding agencies

A **critical characteristic of the US SBIR program is that is does NOT require a match from the small business.** The belief of the US Government is that the success of enough of the projects will result in sufficient additional taxation revenue to more than cover the cost of the programs. These programs have resulted in more than 15,000 firms being established with a total of 50,000 patents awarded and involving about 400,000 scientists and engineers.[4] Some other countries have required matching funding from the small business. **The danger with this approach is that most small firms are cash poor and cannot easily provide a match.** Therefore requiring matching funding is likely to decrease participation in the program.

---

[a] In the United States 'small' is defined as a company with fewer than 500 employees. In Europe, as well as in multinational bodies such as the UN and World Bank, 'small' is defined as fewer than 250 employees.

The main advantage of these types of programs, in our view, is that **they encourage an atmosphere of entrepreneurship among faculty and staff of research institutions** and this leads to a healthy exchange between the academic, research world and the commercial one. The success of the US SBIR programs speaks to the value of the concept.

Other countries, including the United Kingdom, have programs closely modeled on the US SBIR/STTR program. In the UK, university faculty can be the lead PIs on 'Small Business Research Initiative' (SBRI) grant applications, although the majority of grants are given to PIs based in small businesses.

In addition to the advice presented in Chapter 4 on preparing a research proposal such as the importance of significance and impact, there are additional factors that you will need to take into account in preparing an SBIR/STTR proposal.[5,6]

DO focus on your **product** and **not** on the **technology**.

DO include a clear description of the problem your project is meant to solve.

DO indicate why your product will provide a better solution to the problem than currently existing products.

Do provide milestones, including time and money, for bringing your product to the marketplace.

11.8 University–Industry Mobility Grants

In order to further strengthen university–industry knowledge exchange, some funding agencies offer grants that enable the movement of people between the two communities. Programs such as the US National Science Foundation's 'Grant Opportunities for Academic Liaison with Industry' (GOALI) program enable faculty members, post-docs and graduate students to carry out research in an industrial setting, thus gaining valuable experience of real-world industry-based R&D. Alternatively, they allow industrial researchers to spend time on a university campus bringing their practical commercial perspective, insights and know-how to academic research endeavors. Programs to enhance university–industry mobility exist in other countries, for example the UK 'Knowledge Transfer Partnership' program.

As with other proposals involving university–industry partnerships, it is important to address issues such as intellectual property rights, economic impact and dissemination.

DO put appropriate intellectual property arrangements in place.

DO put plans in place for disseminating 'lessons learned' more broadly to your colleagues and community once you or your team members have returned from your partner's 'campus' or industrial site.

DO explain how the engagement will strengthen ongoing interactions between the partners, including laying the foundation for future collaborative research projects.

DO explain how the interaction will strengthen the likelihood of economic impact from your research activities.

DO encourage your industrial visitors to give seminars while on the university campus or participate in other ways with the wider community beyond your own lab.

11.9 'Return to Research' Grants

Some funding agencies offer 'Return to Research' grants which support researchers seeking to resume an research career following a period of absence due to maternity/paternity leave (or other family caring responsibilities), long-term illness, or time in a non-research role. Such grants are typically intended to allow the researcher to work towards being able to submit a more conventional PI proposal.

Some grants of this type are relatively short, perhaps only six months, and support activities such as literature reviews, gathering preliminary data, (re-)familiarization with equipment and techniques, attending conferences and workshops, developing research collaborations, etc. Other grants (e.g., Wellcome Trust 'Career Re-Entry Fellowships') are longer, more substantial awards that include salary and research expenses.

Review criteria for such grants typically focus on the quality of the applicant's track record and career trajectory before the career break, the appropriateness of his or her strategy for returning to research, and the plans to achieve research outputs commensurate with requirements of conventional grant funding mechanisms.

11.10 Summary

There are a number of special kinds of grant competitions organized by funding agencies to promote specific goals. Sometimes these goals arise from the needs of the research community and sometimes from the needs of the society at large. Examples of such grants include:

- Equipment grants designed to increase the availability of modern research equipment.
- Travel grants of various kinds, such as grants funding researchers to spend time in an foreign laboratory or to have a foreign visitor spend time in a local laboratory.
- Grants to support international collaborations.
- Awards to support young investigators.
- Small business awards to encourage the development of research ideas in order to facilitate their advance into the marketplace.

These special awards can amount to significant funding, and are normally reviewed external to the funding agency. The same principles apply to preparing proposals for these special awards as for general research grants. However, there are often additional requirements for these awards. In all cases, in preparing submissions, particularly those involving international activities, explaining how the award may benefit the local research community is extremely important.

Chapter Twelve

Managing the Award

Success is not final, failure is not fatal: it is the courage to continue that counts.

— Winston Churchill

Good news! Your proposal has been successful and you have been awarded a grant. The budget negotiations with the agency are complete and your grant is close to the budget you requested. Congratulations!

However, the realization has begun to sink in that this is not the end of the process. In fact, it is really just the beginning. Now the real work of carrying out the project begins. As the lead researcher, or principal investigator (PI), it is your responsibility to organize and carry out the project within the constraints of time and budget required by the conditions of the award. In other words, **you need to deliver on what you have promised**.

Although this book is primarily about writing a convincing and compelling grant proposal, in this chapter we will briefly outline some of the hurdles you face in carrying out your project. In particular, we will explore the first practical steps you need to take when starting your project, the challenges of managing people and budgets, preparation of reports, managing so-called 'site visits' from the funding agencies and finally looking ahead to preparing for the next round of proposals that could extend the original project in new directions. If you have already thought through the challenges of managing your research program in your proposal, then you will now be in a good position set out a realistic agenda with appropriate scope and ambition, and a practical work-plan.

While this chapter is mostly about what you have to do when you have been awarded the grant, it is actually worth thinking all these issues through (and reading this chapter) before submitting. You should be absolutely sure that you want to sign up for the scale of effort and responsibilities that are involved in managing a grant, especially a large multi-year grant.

12.1 First Steps

Budget

Your first step may be to redo your budget. You may not have got everything you asked for, your research priorities may have changed if there has been progress in your field, and new opportunities with research partners may have emerged.

Don't be disheartened if your budget has been cut by the funders. The funding agency wouldn't have made the award if they didn't think there was something valuable you could do with it. And, as discussed in Chapter Eleven, there are other grant options to supplement or enhance your work.

If your budget has been reduced, it is probably a good idea to work out what can be done with the revised funding amount before you start contacting collaborators, hiring staff or purchasing equipment!

People

Your next steps are fairly clear. You will need to notify your collaborators of the award, and you will also have to confirm appointments for people who are already committed to the project. You will already know how many people you have planned to work on the project and it is likely that you will need to recruit some or all of these. In all likelihood, some of the people you need to recruit will come from outside your home institution. You will need to advertise either electronically or in a printed journal. Alternatively, you may elect to contact colleagues in other institutions for advice and suggestions of promising candidates. Or both! In almost all cases, as discussed in Chapter Nine, recruiting will take longer than you might think, so you need to get started on the process as soon as possible.

> *The budget implies that staff will be on board imme-diately the project begins. This is completely unreal-istic. It takes time to recruit staff, including post docs, so there needs to be a ramp up in the budget.*
>
> **– Quote from a reviewer**

Space

When you submitted your proposal, a representative of the research administration of your home institution signed off on it. The implication of this sign-off is that any matching or other commitments included in the proposal, such as space requirements, (as discussed earlier in Section 4.5) are agreed to by your home institution. In principle, you should also have discussed this with the person responsible for space allocation in your unit (Department or Institute) to identify which particular space could be made available should the proposal be successful.

However, this is not always clarified before a proposal is submitted, because, of course, most proposals are not funded and so the commitments are not required. Therefore, you will need to inform the administration of your institution of your award, and remind them of the space needed for carrying out the project. Space is always jealously protected in universities and other research institutions. Management don't like to see space lying empty. And those who are currently occupying labs and offices won't give them up easily! People need space to work so that before your recruited team member arrives, you will need to have space available for them. If any renovations are needed, this also takes time. All the more reason to start the process immediately!

Equipment

Another item with a potentially long lead-time is the purchase of equip-ment. If you need new equipment to carry out your project, you should look into specifying and ordering this equipment as soon as possible. Your institution may require more than one quote for expensive equipment so you will need to allow time for the bidding process. As discussed in Chapter Nine, as part of preparing the budget for your proposal, you may already have a number of quotes for large pieces of equipment. If this is the case, so much the better, since you are now in a position to begin the purchase process without waiting for bids and thus saving time.

Assigning tasks and responsibilities

Finally, you will need to review your plans with your team, or at least as many of them as are currently on board. Some time will have elapsed since you first generated your proposal and some things may have changed, either because of work you have done or that you are aware has been done elsewhere. In addition, your collaborators may have acquired new commitments (e.g., they may have won their own grants). You will need to assign responsibilities for particular tasks within the project or at least check again on what has already been tentatively agreed. A timescale for completion of tasks and the methods to be used in carrying them out must be agreed upon. You should also discuss priorities with the team and get buy-in, especially if any of the team members have conflicting or even simultaneous tasks.

DO inform your collaborators of the award.

DO start recruiting any new personnel that are needed.

DO start ordering any necessary equipment.

DO inform the administration of your home institution of your award and check on any commitments, especially as regards space.

DO meet with your team and review plans, tasks and priorities.

DO have the team recommit to the program and sign on to the revised agenda.

12.2 Managing People

In many research domains, but especially in science and engineering, projects will typically require teams of people working together to achieve a goal. This will be even more so in multidisciplinary research areas.

Team members will have different skills and abilities (and limitations!) and it is the role of the leader to meld these people into a smoothly functioning team.

"Apparently, he learned his technique from supervising graduate students...."

Although there are a number of different styles used by successful managers, we believe there are a few key concepts for dealing with a team of people:

- **One of the first principles is clear and frequent communication**. Everyone on the team must know his or her responsibility in the project including the time frame expected for completion of sub-tasks.
- **Keeping people on task and avoiding digressions is another important role of the team leader**. One excellent and commonly used approach is to hold regular (typically weekly) team meetings where individuals report on progress and can obtain advice on problems.

This can be extremely useful in solving the inevitable problems that arise in any research project.

- **The leader must set priorities, provide input on unexpected difficulties and guide less experienced members of the team to achieve their intermediate goals**.
- **The leader should also, however, be able to delegate when appropriate** (especially to experienced post docs) and not micromanage. Some of the best ideas and solutions will come from the team and they should have the freedom to show initiative.

If it is difficult for the team to meet regularly, short written reports can be circulated to update all team members about progress. Face-to-face meetings should, however, be held as often as is feasible since this provides the opportunity for exchanging ideas most readily.

Given the rapid advances in internet communication, it is often possible to meet face to face over the web with members of the team who may be located some distance away. This is a great advantage and can be a useful addition to the normal email traffic that occurs between collaborators.

The atmosphere that the leader should aim for in the team meetings is of critical importance. **You should strive to have a very open discussion where people are encouraged to raise and discuss problems that arise**. No one should feel threatened because their results are not turning out as expected. It is far better to have the group face problems and seek solutions together than for some members to feel they must hide what is happening in their particular piece of the project.

Most research teams will include graduate students and sometimes undergraduates. Remember, graduate students are still learning and finding their way, so don't overestimate their initial productivity. One of the responsibilities of the team leader is to make sure that the students are obtaining training in research methodology and skills as well as serving as a useful pair of hands. Provide them the freedom to come up with ideas and solutions to problems that arise, and the confidence to suggest them.

Team meetings can also be used as an opportunity for students to practice the important skill of presenting their work orally. Student presentations should be encouraged. Similarly, students should be encouraged to write reports and circulate these for comments from the other team members.

DO arrange regular meetings for your team, where team members report on progress.

DO maintain a collegial atmosphere in team meetings where people work to solve problems together.

DO be certain that everyone is clear on their particular responsibilities for the project.

DO be sure to set priorities and to review these regularly.

DO provide opportunities for students and postdocs to present their work to the group.

DO NOT assign blame, but stress the search for solutions.

12.3 Managing Budgets

Managing the budget for your project will be one of the critical, but often neglected, tasks in determining the successful outcome. Your proposal laid out a spending plan on a year by year basis, and the funding agency will expect you to adhere to this plan. The amount of flexibility in moving funds between categories will vary from agency to agency but you will generally need to obtain approval before moving a significant percentage of the budget from one category to another.

If your award is similar to the budget in your proposal, your spending plan should be clear. But if your actual award is less than you requested, you may need to modify your work plan. The optimum time to negotiate any changes in the work plan is when you are negotiating the final award with the funding agency.

The key to managing the budget is to keep track of expenditures. Your institution will have an accounting process to assist you in doing so. In general, you will receive reports, typically monthly, about the status of the budget for your award, which should include monthly expenditures and the total amount remaining. In addition, it is also useful to keep track, at least approximately, of the expenditures from your award independently. For one thing, institutions have many awards to keep track of and it is not unknown for errors to be made, and for expenditures to be charged to your account that were not incurred by your award. **You should check the monthly reports from the budget office of your institution** and see that they agree with the expectations from your independent records. If not, you will need to track down the problem. Perhaps a piece of equipment has been charged incorrectly, or perhaps someone else's graduate student's stipend is being charged to your account. The sooner these errors are found the easier they are to correct.

Travel is often another sensitive item in a budget. There are usually rules about travel and *per diem* expenses both from the agency and from your home institution. You will need to be careful that **your expenses are consistent with these rules** or you or your colleagues may not be reimbursed for your travel expenses. Travel expenses have also, on occasions, been a source of embarrassment. For example, conferences are often held in exotic and expensive locations. Sending too many people to such conferences, while permissible, may not look like sensible expenditures of grant funds if made public. Remember the 'red face test'. How would you feel if you saw the information prominently displayed in the local newspaper or television? Can you justify the expense in terms of the added value to your research?

The timing of expenditures is sometimes an issue. As discussed in Chapter Nine, expenditure will vary over the lifetime of a grant. If the funding is released by the agency on a year by year basis, rather than as a lump sum, there will usually be a report required on the first year's expenditures before the second year's funds are released. If you had spread the funding fairly equally over the total number of years in your original proposal, **you may find that you are grossly under-spent in the first year** either because you did not have all the personnel in place for the full year, or possibly because equipment purchases have been delayed. This is an issue that is better anticipated in advance, if possible. However most agencies are understanding of these issues and will generally forward your second year's funding as needed. The ability to estimate funding by year is even more of an issue for early career researchers who are building a team from scratch or for researchers in systems where the funding available for research has increased dramatically and quickly.

A similar problem can be encountered at the end of the award period. A slow start can also mean that **funds remain at the nominal end of the award and there is still work to be done to complete the project**. Most agencies have a standard grace period (sometimes termed a 'no-cost-extension') that allows expenditures for some time (usually a few months but possibly up to one year) after the nominal end of the award. Beyond that time, the PI must usually provide a good explanation for continued expenditures and at some point the agency is very likely to pull back any remaining funds.

An even more critical timing problem can be encountered towards the end of the grant period. If you are spending the award at a rate that will **overspend the total grant before the end of the award period** this may mean that you will have to lay people off or make some other drastic budget adjustment. Laying off people may involve contractual problems and, at a minimum, will be disruptive and should be avoided. In addition, your reputation as a grant manager will suffer, and you may lose good team members whom you may want to continue working for you during the next grant. Looking ahead to the next grant should always be on your mind (see Section 12.5).

In any case, most institutions will not approve of overspending on grants and will normally **require that any overdrawn funds to be repaid**. This will usually involve a penalty of some kind, and should definitely be avoided.

The lesson about budgets is fairly clear. **Keep track of what you and your team are spending from the award and make sure this is consistent with the overall budget**. If you need help managing your budget, the finance office of your home institution would be one place to consult. Remember that adjusting expenditures early is much easier than trying to do so towards the end of the award period.

DO keep track of expenditures and make sure these are consistent with the budget.

DO check the timing of expenditures especially at the beginning and towards the end of the award period.

DO make any needed adjustments to expenditures as early as possible.

DO NOT (**under any circumstances whatsoever!**) overspend the budget of the award.

12.4 Progress Reports and Site Visits

Most funding agencies will require a **written report at the end of the award**, listing the achievements of the project including publications and presentations at meetings. If the award is for more than a year or two, there will likely be the requirement of intermediate reports at the end of each year or sometimes more frequently. Normally these written reports do not provide any difficulties and give the PI an opportunity to review the status of the project and to highlight any significant achievements. It may be a good idea to update a draft version of the report on a regular basis — adding significant outputs/outcomes as they arise, rather than leaving them until shortly before the report submission date. Updating outputs and outcomes could be a regular agenda item during team meetings (as discussed in Section 12.2).

If there are any unexpected results which might lead to new directions for the research, these can also be mentioned. Don't be afraid to follow more interesting or more promising results if they arise. Similarly, **you should be willing to close down sub-projects if they are not working**. Let your funding agency know what you are doing, but the agency staff will most likely want you to pursue the most promising opportunities.

Funding agencies are often asked by the government to provide information on progress. Written reports provide useful input to respond to these requests. If the PI believes that significant new results have been obtained, this should also be transmitted to the funding agency. Government funding agencies like to know of any 'gems' that can have an impact either on the science or on society. Breakthrough results can help the agency solicit new funding which will help all the researchers in the field. However, you must make sure that any unusual or surprising results are, first of all, correct and secondly that they are truly significant before sharing them with the agency.

Staff from the funding agency will sometimes visit the site of a project to learn firsthand how the project is progressing. **Such visits should be welcomed as they provide you with the opportunity to speak directly with the staff about the aims and progress of your project** as well its value in training students. The PI should make sure that as many participants as possible are present during the visit so that they can be available to answer questions and speak informally (and hopefully enthusiastically) about the progress of their particular part of the overall project.

If a project is complicated, costly and extends over a number of years, the funding of the later years may depend on progress. In such a case, the agency may decide to carry out an intermediate site visit, even in some cases using peer reviewers to assess progress on the award. Since the remaining funding may depend on performing well at a site review, it is important **to plan the presentations carefully and to practice them beforehand**. Usually it is advantageous to have collaborators, whether from other research institutions or from industry, present at the site visit, even if they are not normally on site, since this helps to indicate their commitment to the project. The same is also true for a senior member of the administration of the home institution. Generally, it is a good idea to allow the people most directly involved in particular areas of the project to make presentations since they are most familiar with the work and can therefore more easily respond to questions. However, the PI should also be prepared to give an overall summary of the progress of the project and put the various segments in context. If problems have arisen, which is often the case in research at the cutting edge, these should be stated and the strategy for overcoming the problem should be presented clearly. Reviewers will appreciate honesty and may even be able to suggest solutions.

Site visits also provide an opportunity for graduate students to present material to a scientific audience. However, particularly with new students, these presentations should be carefully practised and critiqued beforehand, both for the sake of the students and the project.

DO submit progress reports on time.

DO treat site visits from agency staff as an opportunity to convey the excitement of the project.

DO prepare and rehearse any formal presentations to visiting review groups and/or agency staff.

DO keep agency staff informed about interesting new results.

12.5 Preparing for the Next Round of Proposals

Once the current project is beyond the half way mark, you might begin to think about submitting a new proposal to the next appropriate competition.

This can take the form of a continuation of the direction of the current proposal, or branching out in a new direction, possibly driven by discoveries or at least suggestions derived from the current project. This can be particularly important if you want to retain key team members whose salaries come from the current funding. They will need to know the funding for their jobs is in place or they will (if they are sensible) start looking for other positions.

It is always important to keep questioning the value of the specific research question you are asking in any proposal. Therefore, you should be cautious about settling for just 'more of the same'. As discussed in Chapter Two, you must continue to ask questions like: *What new information will be obtained? How will my new project advance understanding in the field?* Sometimes an unexpected discovery can lead to exciting new directions and you should always be open to such possibilities.

One critically important factor in thinking about a new submission is to **make certain that you have significant outputs from your current grant** (see Chapter 7). These outputs can be papers published in peer reviewed journals, invited talks at national or international conferences, or even patents, if this is appropriate. The number of students graduated and postdoctoral fellows trained are other important 'person-power' outputs that can be important for a country trying to build its research workforce. However, it is important to stress that your work will be judged mainly based on peer reviewed publications.

One of the questions the funding agency will ask, and certainly a knowledgeable reviewer will ask, is: "**What publications have come from previously funded work of this person or group and what impact have they had?**" Therefore it is extremely important to set a priority to publish as much of the current work as possible in as high impact journals as possible. This will be of critical importance in obtaining funding for a new project.

Presenting your work at conferences or professional meetings and even at universities or research laboratories is also a way to help your work to become known. However, the most valuable currency remains publications in peer reviewed, high impact journals.

As your group pursues the current project, there may be opportunities to **obtain some preliminary data related to a new direction** that may lead to a new project. As discussed in Section 2.5, it is increasingly expected, especially in biological and medical sciences, that preliminary data should be provided to support any new idea that is being proposed. So if you suspect that you may want to pursue a particular direction in a new proposal, it is advantageous to obtain some data in support of your idea during the course of the current project. It is reasonable to use some funds from the current grant for this purpose, provided that the new direction falls broadly within the scope of the original proposal.

The only final caution is to make sure that preparing for a new proposal does not distract your team from completing the work which is currently funded.

DO think ahead to the next grant competition.

DO make sure that you have significant outputs from your current grant.

DO generate preliminary data for a new proposal if at all possible.

DO NOT necessarily continue on the same research path but look for interesting new directions.

12.6 Summary

Once you learn that your proposal has been funded there are a number of things you need to do to ensure the success of your project:

- **DO** inform your collaborators and your institution's administration about your grant.
- **DO** begin recruiting any new people that the project requires and order needed equipment as soon as possible.
- **DO** arrange regular team meetings of the group to update everyone on the team on progress and to discuss any difficulties that arise.
- **DO** make sure that all the people on the team know their tasks and the timetable for completion.
- **DO** keep track of budget expenditures and check against institutional reports on a regular basis.

- **DO** provide progress reports as requested and prepare carefully for any site visits requested by the funding agency.
- **DO**, if possible, obtain preliminary data related to new ideas for a potential next proposal.
- **DO** publish completed work in high impact journals as quickly as possible.

Appendix One

Organizing a Research Proposal Competition

Setting up an agency to disperse funds for research requires many, many processes centered around personnel, budgets and oversight of the agency, even after the vision and the goals of the agency and the use of the funding have been agreed upon by the government. There are many important questions related to all these areas but we do not propose to discuss these in any detail here. Useful models are available around the world and in most cases concepts used elsewhere can be adapted to local use.

As in most human activities, leadership is critical, and the choice of the Director is the most important decision in setting up a new research funding agency in a country. While knowledge of the local science culture is important, ideally we believe that **international experience especially the experience of operating in a competitive, high quality science culture is also extremely useful.** If the agency can find a citizen who has had international experience and is willing to return home to lead the agency, this would be of immense help in establishing the right spirit in the agency.

The next most important issue is the quality of the staff of the agency, particularly the professional staff. We believe that there is great value in having research scientists serve in the funding agency although this can lead to a difficult recruiting process. The US National Science Foundation has a strategy of using **rotators**, i.e., researchers normally working in universities, who spend one or two years at the agency and then return to their academic posts. This has the great advantage of bringing needed expertise into the agency, although it can sometimes be difficult to find willing rotators. The knowledge and insights of such active researchers can be extremely helpful, both in evaluating proposals and in setting strategic directions for the agency. Professional staff understand the research and educational 'culture' which exists in universities which can be very useful in dealing with reviewers.

In this appendix, we plan to deal with only one particular activity of the funding agency, albeit a critical one — how to set up and run a peer reviewed competition for research proposals. This is a central activity of all funding agencies and knowledge about the potential pitfalls of such a process can be useful both to the staff of the agency and the researchers who are seeking funding from the competition. We will provide suggestions for the degree of focus of any competition, the language used, the selection of reviewers and the mechanics of the reviewing as well as the content of the call for proposals (CFP).

A1.1 How Focused should the Competition Be?

To some extent, this will depend on the mandate given to the funding agency by the government of the country. For a large country with a well developed scientific and technical workforce this mandate can be quite broad and cover all, or most, science and engineering fields. However smaller countries, especially those that do not have a long history of research funding from the government, are wise to be more restrictive in their choice of areas to fund. This can be difficult because it means setting priorities that are appropriate to the country. The scientists themselves will find this difficult because they will all want to have their own field receive a high priority because they believe it is the most important. So the priority setting needs to be done by some group that is more independent and has the overall good of the country at heart.

Even though this is difficult, we believe that it is extremely important if the funding from the government is to make a difference. If no priorities are set and the funding agency attempts to fund all fields of academic work, the danger is that the funding will be spread very broadly and no one area will have sufficient funds to really prosper and become more recognized internationally.

We believe that Ireland provides a good example of this priority setting approach. Before Science Foundation Ireland (SFI) was set up, the Irish government commissioned a study of priorities and with the aid of this study decided that the new funding agency, SFI, would only concentrate on two areas Biotechnology (BIO) and Information and Communication Technology (ITC). Both of these areas were interpreted quite broadly to include a good deal of solid state physics, material science, computer science and most branches of biology — but not clinical studies which were already funded under the mandate of the Health Research Board. In practice this meant that the bulk of the funding went to these two areas and a small amount (about 5% or less) went to small grants in all other science areas such as geological sciences, astrophysics and mathematics. Since then SFI has added Energy to its other two priority areas. The fact that Ireland concentrated its science funding initially in these two areas has meant that they have now made a significant leap forward in terms of international recognition.

In addition to spreading the limited funding too thinly, a lack of focus also makes the reviewing of proposals more difficult. If there are no restrictions in a particular competition then proposals will be submitted in a very wide range of areas, and there will be fewer proposals in any one area, making it difficult to find reviewers who can review a significant number of proposals in any given competition.

If the focus of the agency is on national objectives, one must ask how the research is linked to on-the-ground outcomes that benefit the nation concerned. Those benefits can usually be delivered by the industrial base of the nation. This achieves two outcomes, *viz.* linking the research culture to the industries that will deliver the outcomes, and strengthening the industrial base of the nation concerned. Most successful research funding schemes across the world achieve these outcomes by having a significant part of the funding dedicated to 'industrial partnerships' where the industry partner provides some funding and involves industry staff in the proposed research. Given the move towards more global industrial development, the partner industry can be a multi-national company provided the company has a sustainable base in the nation concerned.

Therefore, in spite of the difficulties of setting priorities, for countries that are still trying to build a research infrastructure, **we strongly recommend that a small group of areas be selected where funding is concentrated**.

A1.2 What Language to Require for Proposals?

This becomes an issue for those countries that do not have English as a native language and which are not large enough or do not have a well developed science research community, so it would be difficult to find enough reviewers from within the country who could review proposals without any danger of bias, either positive or negative. Countries like Ireland or Australia which do not have a very large science or engineering research community have the advantage of using English as the common language, so it is comparatively easy for them to use researchers from other English speaking countries as reviewers. On the other hand, a number of EU countries where English is not the native language such as France, Spain, Italy and Germany can use their own language for proposals because they have large science communities from which to select reviewers. Nevertheless, most EU countries, and the EU itself, elect to have proposals submitted and reviewed in English.

In many cases, small countries choose to require that proposals are submitted in English since this has become the universal science language in almost all parts of the world. Nearly all international conferences use English as the common language. If a scientist is unable to read English this limits his or her access to the scientific literature enormously.

On the other hand, choosing to use English can have problems for the authors of proposals and special care must be taken to ensure that there are facilities available to assist authors and make sure that their proposals are presented in clearly understandable and grammatical English. In some cases the funding agency may have to provide facilities to ensure this. In other cases the research institutions should be required to do so.

A.1.3 What Kind of Reviewers to Use?

The simplest case is where the country and its scientific community are large enough that internal reviewers from within the country can be used to review all proposals. Some precautions are usually taken to ensure that there is not some conflict of interest in specific cases, where people may have worked together in the recent past or have some other relationship that makes either a real conflict of interest or, what is at least as important, the appearance of a conflict of interest possible. One advantage of using reviewers from within the country is that there are usually less travel expenses involved, and, normally, such **reviewers are not compensated financially since it is understood that this is an obligation to the research organization within the country.** Reviewers will sometimes themselves be authors of proposals and so benefit from the funding within the country.

However if the research establishment in the country is not very large, it is difficult to avoid at least the appearance of a conflict of interest if reviewers are chosen only from the home country. People in similar fields know one another, and have often established friendships (or rivalries) which make it **difficult to avoid at least the appearance of a conflict of interest** even if not the reality. A robust peer review system requires that the proposals be reviewed objectively and without any prejudice, otherwise the credibility of the agency will be called into question. Authors of proposals that are rejected can always claim that the reviewers were biased against their work and for a small group of researchers reviewing each other, this can be true in some cases.

Therefore, in these countries, we strongly recommend using reviewers from outside the country either exclusively or at least in the majority of cases. This avoids, to a very large extent, the problem of conflict of interest, although it is normally a good idea to still require a statement from reviewers that they are not conflicted with the authors of any of the proposals they review. However, the chances of such conflicts are much reduced for reviewers outside the country where the proposals originate.

The other great advantage of using reviewers from outside the country is that it **provides the opportunity to select reviewers from a much wider pool of candidates** and thus to be able to match reviewers to particular proposals much more closely. In trying to improve the quality of proposals within a country, the use of expert international reviewers also means that the normal international standards will be used in evaluating proposals and this will have the effect, over time, of improving the quality of the submitted proposals. At the very least, it will provide input from reviewers who will apply international quality standards. Their comments will be shared with the authors of the proposals, provided that the reviews are distributed (anonymously) to the specific authors of each proposal.

Such sharing of anonymous reviewers comments is a common international practice, which we strongly recommend.

Another advantage that can sometimes occur with the use of 'out-of-country' reviewers is that such reviewers can often identify international centers closely linked to the subject of a proposal and suggest that the investigators establish links to those centers. These links can be important in developing the scientific capability of a small country.

A question that might be raised is whether any local (from within the country) reviewers should be used? Generally we would recommend against it. **Even a single local reviewer will often be perceived as having enormous influence over the review process whether or not it is true.** The disadvantage of the appearance of conflict of interest outweighs any perceived advantage. At SFI, in a particular competition, we used panels of about a dozen reviewers for different fields. In order to help convey the high quality and objectivity of the review process to the local community, SFI elected initially to have one local member on each panel. However, the community believed that this one member out of the dozen had extreme influence over the remaining members of the panel and therefore was a source of bias. SFI discontinued the practice after a couple of rounds of the competition and settled on all external reviewers.

The main **disadvantage** of using external reviewers is the additional cost and administrative trouble for the funding agency staff. It is probably harder to find suitable external reviewers since these are not so well known to the staff. Plus, travel costs need to be paid if they are to be used in a panel. Even if the reviews are carried out by mail, or more likely over the internet, the external reviewers have to be paid for reviewing proposals, whereas the local reviewers could be expected to do so for free since they in turn benefit from funding from the agency. In addition, external reviewers are generally less familiar with the budget situation in a foreign country especially personnel funding such as the need to pay summer salaries or not, and the typical process for paying graduate students. However, these latter are not serious constraints and we believe that the advantages of using predominantly, if not exclusively, external reviewers in countries that do not have a large enough scientific infrastructure is well worth the extra trouble and expense.

If reviewers external to the country are to be used, this almost certainly requires that the proposals be submitted in English. While this does add an extra burden to the authors of the proposals, at least in science and engineering, this burden should not be extreme since at least a reading knowledge of English is required to keep up with current literature in most science and engineering fields. However, as mentioned in Section A1.2, the funding agency may have to take precautions to provide translation help as needed or at least to encourage the research institutions to do so.

A1.4 How should Reviews be Carried Out?

The use of the internet has greatly increased the ease with which reviewers can be contacted, proposals can be circulated and reviews submitted. There are now many companies that provide commercial programs[1] to manage the reviewing of proposals and papers electronically[a]. However there are a number of options about what kind of reviews are most useful in any given competition and they each have their strengths and weaknesses.

A1.4.1 *Postal or mail reviews*

Perhaps a better title for such reviews would be 'individual reviews'. In many ways, this is the simplest of the review methods. Basically, a review is prepared by a single reviewer and submitted to the agency. In most cases, the reviewer can access the proposal online and after reading the proposal can submit her or his review online too.

In some cases, the agency will request that the review be submitted under various headings with separate scores for each heading. There is also normally the requirement for a total numerical score for the proposal.

Our view is that, while using subheadings for reviews is sometimes useful, what is more important is to make sure that the reviewer has the opportunity to provide an overall assessment of the proposal. In some cases, especially using numerical ratings, it is possible for scores for individual

---

[a]The US National Science Foundation developed the program Fastlane internally and has used it successfully for many years.

sections to be high but there can still be a fatal flaw in the proposal that makes it a very poor candidate for funding. For example, the proposal might be clearly written, the problem might be important but a reviewer may be aware that the work has already been done and published by someone else. Alternatively, the reviewer may find a technical problem that will make it impossible to carry out the project as described even though the proposal itself is well written, and deals with an important topic.

Numerical ratings provide a useful summary, but since different reviewers have different standards for such ratings, **it is important for the staff of the agency to read the actual written reviews carefully**. This often gives a much better idea of what the reviewer really thinks about the proposal rather than just relying on a single number to rank it.

Normally at least three postal reviews should be required for assessment of any one proposal. However, in the case of large complex proposals with many different aspects it may be necessary to have as many as ten or more postal reviews from different people to completely assess it. Given the different approaches of the different reviewers, this makes the careful reading of the separate reviews all the more important.

One disadvantage of postal reviews is that in some cases reviewers, who may have agreed to provide a review, do not actually send in the review in time before a decision has to be made. Therefore, it is often a good idea, if a minimum of three reviews is needed, to actually ask more than three reviewers to provide reviews, making it more likely that at least three will be received by the deadline.

A.1.4.2 *Panel reviews*

The next most complex review process is to use a panel of reviewers who meet together, usually at the agency headquarters, to discuss and rate a group of proposals. The common practice is to have individual mail reviews available to the panel reviewers some time before the meeting so that the panelists will have had a chance to read them all. We believe there are advantages in having the panel members also serve as postal reviewers. One reason is that this more or less guarantees that the postal reviews will be received at the very latest by the time of the panel meeting, so that it is not necessary therefore to request extra postal reviews.

The main advantage of a panel meeting is that the proposals and the reviews are discussed by a group of experts in the general area of science. This almost always leads to a more reliable assessment of the proposals, including the relative merits of the different proposals and their ranking. The panel can be charged with ranking the proposals or at least sorting them into groups such as **Must fund**; **Maybe fund**; and **Do not fund**. The panel can also be asked to provide a brief summary of the reasons for their decision concerning each individual proposal and this can be shared with the authors of the proposal. This provides very useful feedback both for successful and, especially, unsuccessful proposals, and should improve the quality of the proposals over time.

One question that arises is who should chair the panel meeting? We suggest that the best chair is one of the panelists, preselected before the meeting by the agency. The alternative is to have the panel meeting chaired by a staff person from the agency. We certainly agree that a staff member must be present throughout the panel meeting both to answer questions and to listen to the discussion during the meeting. However, except in unusual situations, the panel will normally respond better if they feel that the panel is autonomous by having a chair from its own membership and is thus in a better position to provide independent advice to the agency. It is extremely useful to the staff of the agency to hear the discussion of the panel members, since this allows them to understand clearly the reason for the panels' decisions and be in a better position to respond to questions both from other agency staff and from authors of proposals who might ask for further information.

A.1.4.3 *Site visits and reverse site visits*

For even more complex and expensive proposals, the agency may decide to carry out a site visit, using a team of reviewers to visit the site where the majority of the project is to be carried out. Alternatively it may ask the principal investigator (PI) to present the proposal before a panel of reviewers and answer questions (reverse site visit).

The former is particularly useful if complex equipment or facilities are required for the project. This provides the review group the opportunity to see the state of readiness of the equipment and facilities. Visiting the site

Table A.1 gives a brief summary of the advantages and disadvantages of the different review methods.

Table A1. Summary of different review methods.

Methods	Advantages	Disadvantages
Postal	Least expensive Simple to administer	No interaction between reviewers No opportunity to question PIs May need to have extra reviewers to ensure enough reviews on time
Panel	Reviewers interact with each other More reliable rating of proposals Can guarantee enough reviews are received on time	Travel costs for reviewers More complex to administer
Reverse site visit	Reviewers can question PI Reviewers interact with each other	Travel costs for reviewers More complex to administer More effort from the PIs
Site visit	Reviewers can question PI and other participants Reviewers see facilities and equipment Reviewers can assess commitment of collaborators and administration Reviewers interact with each other.	Extra travel costs for reviewers and agency staff. Much more complex to administer Much more effort for PIs and the rest of the team

of the project also gives the reviewers the opportunity to meet more of the participants, including any collaborators and the administration of the institution, giving them a chance to gauge the degree of commitment of these groups to the project. On the other hand, there is significant extra cost and administrative work involved in organizing a site visit, especially if there are a number of competing proposals since all would have to be visited to ensure fairness.

In such cases, a reverse site visit may be a useful option, where the PIs of competing proposals present their projects before the same review team at a single neutral location, often the headquarters of the funding agency. This still provides the opportunity for the reviewers to interact with the PI, and to ask questions about particular aspects of the proposal that may have arisen either in the postal reviews or in the PI's presentation. **It is surprising**

how much extra information can be gained from direct interaction with the PI as opposed to simply relying on the written proposal. This is often extremely valuable for a complex proposal where the leadership qualities of the PI can play a major role in the success of the project.

Basically, the more complex the proposal, the more thorough the review needs to be, even if the cost of the review and the administrative overhead increases. Thus simple proposals can be reviewed either only by postal reviews or panel reviews and more complex proposals may require site visits or reverse site visits.

A.1.5 The Use of Pre-Proposals

There are various views on the value of requiring a brief pre-proposal instead of an immediate call for a full proposal. The advantage of a pre-proposal stage is that it requires less effort both on the part of the agency staff in reviewing the pre-proposals and particularly on the part of the applicants in preparing the pre-proposals. It also provides the opportunity for some feedback to applicants to improve their full proposal submissions.

A pre-proposal is used most often when a very large number of proposals is anticipated, and the effort to review them all is daunting, requiring many reviewers. For example, if a full proposal has a length of 15 pages in the body of the proposal plus CVs, budget, and letters of support from collaborators, it is probably unrealistic to expect any single reviewer to review more than 8 to 12 proposals. On the other hand, if a pre-proposal is only three pages long then a reviewer could be asked to review as many as 30 pre-proposals reducing the number of reviewers by about a factor of three. The pre-proposal stage could therefore reduce the number of full proposals by a factor of three or even four resulting in a net reduction in the number of reviewers needed. Plus, the applicants only are required to prepare full proposals when there is a reasonably high success rate to be expected.

But there are downsides to such a scheme. First, two competitions are required before a funding decision is finally made. This complicates the process and certainly lengthens it. Second, some reviewers believe that it is difficult, if not impossible, to evaluate a pre-proposal accurately because of the paucity of information that can be provided. Most reviewers would agree that decisions on pre-proposals are likely to be less accurate than decisions on a full proposal, where more detail on all aspects of the proposal can be presented. However, this is not a universal view among reviewers, some of whom believe that the essence of the research idea can be presented briefly but clearly.

Perhaps the main lesson is that the agency should always carefully consider the length requirements for any proposal and keep it to the minimum that will still allow sufficient detail to be presented to reviewers for them to make a reasonably accurate assessment.

A.1.6 What Information should the Call for Proposals Provide?

When a funding agency decides to run a research competition, the agency will circulate and post on the agency's website a document referred to as the CFP. This document should provide the information that authors need to know in order to prepare a proposal. Important information that the CFP should contain includes: The submission deadline for the competition; the eligibility criteria for who may submit proposals; the evaluation criteria and any weighting of these criteria; the instructions for the content of the actual proposal including any space limitations; information on how much detail is required in the budget; and instructions on submission of required curriculum vitae. The agency may also wish to inform the authors of proposals about how the proposals will be evaluated. We comment on a few of these points in the following subsections.

A1.6.1 *Deadlines*

With web based or internet submission of proposals, it is now comparatively simple to apply hard stops on the time of submission of proposals. This avoids the need for agency staff to return proposals that have been submitted after the deadline.

We have always been in favor of firmly enforced deadlines since we believe this is the only way to ensure a level playing field for all participants. Once a proposal can be submitted even one minute late, then why not 5 minutes, why not 30 minutes and so on? The only way to avoid this issue is to state very clearly the actual deadline (e.g., 1700 hours on Monday, April 8, 2013) and stay with it. The only exception is if there is a problem at the agency that prevents some participants from being able to submit their proposals on time. Problems at the individual research institutions should not be an excuse. Authors will quickly learn that it is important to prepare their proposals for submission some time before the deadline to be sure that they can deal with any last minute difficulties.

Once a firm deadline has been enforced on a particular competition, you can be sure that there will be much more care taken in future competitions by both the authors and the research institutions to make sure that proposals are submitted on time.

A1.6.2 *Evaluation criteria*

The evaluation criteria used to evaluate the proposals should be spelled out in detail in the CFP. These criteria should also be made clear to all the reviewers of the proposals. The agency may elect to provide some priority weighting to particular criteria but this is rather unusual and such prioritization is normally left to the reviewers based on their own experience.

Some agencies rely on rather general criteria, but others are more specific. For example, the US National Science Foundation uses basically two criteria for almost all its competitions *viz.* (a) The intrinsic scientific merit of the proposal and (b) the potential impact on the field or on neighboring scientific fields. SFI uses the following review criteria in the review of what are called Principle Investigator Awards:

- Quality, significance and relevance of proposed research.
- Quality of PI's *recent* research track record.
- Quality of proposal's potential to underpin Biotech/ICT.

The last of these criteria is used because of the initial focus of SFI on Biotech and ICT.

If there is to be any special emphasis on cooperation with industry or other practical outcomes then this should be clearly spelled out in the CFP.

A1.6.3 *Information to require in the body of the proposal*

We suggest that the body of the proposal must answer the following simple questions:

(a) **What question does this proposal intend to answer?**

Many proposals fall down in failing to address this point. Instead of stating a crisp question that the proposal intends to answer, the authors will write about 'studying a topic' or 'probing further into an issue'. Science is generally hypothesis driven and it is critical to ask a specific question that can move the field forward. While in some fields more general exploratory questions are accepted, in biological sciences reviewers will be unforgiving if they do not see a clearly formulated hypothesis.

(b) **Why is this question important?**

This section provides the opportunity for the authors to present the context of their proposal. What previous information on the topic is available? This is the place where the current literature can be reviewed. If the work described in the proposal is successful, how will this move the field forward? If the proposal will result in a cure for cancer, this should be stated here. Or if there is expected to be any other impacts either economic, social or educational, these should be spelled out clearly. It is important here for the authors to demonstrate that they are familiar with previous work on the topic and to explain how their proposal will add substantive new and important information.

(c) **How will this question be answered?**

In many ways this section is the key to the whole proposal. In it the authors need to describe in as much detail as they can, precisely how they will carry out the work. What problems do they anticipate? How will they deal with these problems? What is the evidence that their approach will work? Is there new technology or instrumentation required? Informed reviewers will scan this section carefully to make sure that in their opinion the authors are likely to be able to solve the technical problems and successfully carry out the measurements that they propose and that these measurements will indeed answer the question posed in Section (a). It therefore behoves the authors to make sure that their description of their approach is sufficiently detailed so as to convince the reviewers that the proposal will indeed succeed.

(d) **What evidence is there that the authors are capable of carrying out the work described in the proposal?**

In this section the authors must demonstrate that — using the information contained in their CVs — their previous record indicates that they have the experience and competence to carry out the work proposed. Normally this would involve similar work carried out successfully. However, if an author is entering a new field or subfield, prior success in another area can also be an excellent indicator of likely success in a new area. This is also the place where the specific expertise of collaborators, including industrial collaborators, and their explicit contributions to the project should be described.

A.1.6.4 *How long should the proposal be?*

The length of the proposal is normally determined by the amount of the award. If an award is large, more work can be expected from more people and as a result the description in the proposal needs to be more extensive and requires more space. On the other hand, if the award is comparatively small, for example less than about US$300,000 over approximately three years, then fewer people would be involved and the project would be simpler to describe and not take up so much space.

Generally it is useful to limit the length of the proposal, since this means less work both for the reviewers and for the authors of the proposals. On the other hand enough space must be permitted to allow the authors to explain their project in enough detail to be useful for the reviewers. Normally we believe that brevity should be encouraged. For example for the 'small' awards mentioned above, we believe that a total allowed length of ten pages for the body of the proposal is probably sufficient, and encourages the authors not to introduce extraneous material into the proposal.

Space limitations are a little harder to enforce electronically than deadlines, and depend on how the proposals are actually submitted. Should proposals that exceed the limit be simply rejected and returned without review? Or should the agency remove any pages over the limit and send the abbreviated proposal out for review? This will have the effect of annoying the reviewers and will therefore penalize the authors. The funding agency should make clear in the CFP which option will be used thus reminding the authors to be careful in crafting their proposal.

Another technique that authors sometimes use to sidestep any space limitation is to use a small font size. This again has the effect of annoying the reviewers and penalizing the authors. However a better approach is to specify the font size as well as the spacing in the CFP.

A.1.6.5 *The budget*

The agency should provide some indication of the expected size of the award in the CFP since this will provide information on how complex a project will be considered. If awards of different sizes will be made, the range needs to be specified, particularly the maximum award that will be allowed.

An important question is how much detail to require in the budget presented in the proposal? Our general advice is not to require too much. Usually a breakdown into four or five broad categories by year of the award, such as Personnel, Equipment, Travel, Supplies & Services and Overhead, is probably sufficient. First, because the reviewers may not be familiar with the details of personnel pay in the country, it may be difficult for them to comment usefully on salaries and stipends. Second, even if the proposal is funded, the award may be less than requested so that the actual spending may differ from the original budget. Thus requiring a great deal of detail in the proposal budget may not be very useful.

Nevertheless, it is valuable to request information on large equipment purchases, even to require vendor quotes if possible. Travel is often another sensitive issue and international travel in particular should be justified in the budget. Overhead paid to the research institution is usually negotiated by the funding agency with the research institutions to allow for infrastructure costs of the research. Commonly, this negotiation is conducted every few years.

A.1.7 Educating the Research Community

One of the main reasons that a country funds scientific and technical research is to improve the quality of the research being carried out in the country, and therefore to increase the international reputation and visibility of this research. This will also improve the quality of the researchers, particularly young researchers who are being trained within the country. The hope is that this will in turn lead to more indigenous technical spinoffs as well as increasing the likelihood that international companies may choose to carry out research and development in the country.

Therefore one of the main responsibilities of the funding agency is to use the funding to improve the quality of the research. One method of improving quality is to share the reviews of the proposals carried out by international reviewers with the authors of the proposals. These reviews are normally shared anonymously in order to preserve the privacy of the reviewers and to encourage honest and open criticism.

Another useful approach is to provide general feedback about proposals previously submitted and reviewed by sending agency staff to the research institutions to provide information on the process plus general features of the reviews and to respond to questions from the researchers. Explaining the details of new CFPs can also be carried out by agency staff. The quality of these sessions will be influenced by the quality of the agency staff and provides additional reasons to seek qualified science professionals in these roles.

Appendix Two

General Advice/Guidance on Grant Writing: Links

United States

- NSF:
  - A Guide for Proposal Writing: http://www.nsf.gov/pubs/2004/nsf 04016/nsf04016.pdf
- NIH:
  - Grants Process Overview: http://grants.nih.gov/grants/grants_process.htm
  - Grant Writing Tip Sheets: http://grants.nih.gov/grants/grant_tips.htm
- NIH — NIAID:
  - Writing a Great Grant Application — Questions and Answers: http://www.niaid.nih.gov/researchfunding/qa/pages/writing.aspx
  - Ten Steps to a Winning R01 Application: http://www.niaid.nih.gov/researchfunding/grant/strategy/Pages/stepswin.aspx
  - Pick a Research Topic: http://www.niaid.nih.gov/researchfunding/grant/strategy/Pages/2picktopic.aspx
- NIH — NINDS:
  - How to Write a Research Project Grant Application: http://www.ninds.nih.gov/funding/write_grant_doc.htm
- SSRC:
  - On the Art of Writing Proposals: http://www.ssrc.org/workspace/images/crm/new_publication_3/%7B7a9cb4f4-815f-de11-bd80-001cc477ec70%7D.pdf

- ARPA-E
  - ○ 5 Deadly Sins and 5 Best Practices of Proposal Writing: Views of Two ARPA-E Program Directors: http://arpa-e.energy.gov/sites/default/files/documents/files/Deadly%20Sins%20and%20Best%20Practices%20-%20ARPA-E%20University%20-%20October%203%202012.pdf

United Kingdom

- EPSRC:
  - ○ Preparing a Proposal: http://www.epsrc.ac.uk/funding/howtoapply/preparing/
- BBSRC:
  - ○ Research Grants Guide: http://www.bbsrc.ac.uk/web/FILES/Guidelines/grants-guide.pdf
- ESRC:
  - ○ http://www.esrc.ac.uk/funding-and-guidance/applicants/how-to.aspx
- NERC:
  - ○ Handbook and Referee Guidance Materials: http://www.nerc.ac.uk/funding/application/howtoapply/forms/
- MRC:
  - ○ Guidance for Applicants and Award Holders: http://www.mrc.ac.uk/documents/pdf/guidance-for-applicants-and-award-holders/
  - ○ Reviewers Handbook: http://www.mrc.ac.uk/documents/pdf/reviewers-handbook/
- Wellcome Trust
  - ○ http://www.wellcome.ac.uk/Funding/Biomedical-science/Application-information/wtvm052727.htm

Other International

- The Foundation Center
  - ○ Proposal Writing Short Course: http://foundationcenter.org/getstarted/tutorials/shortcourse/index.html

- Human Frontier Science Program
  - Art of Grantsmanship: http://www.hfsp.org/funding/art-grantsmanship

Selected Developing Economies

- South African Medical Research Council
  - Writing a Research Grant Proposal: http://www.mrc.ac.za/research-development/researchgrant.pdf

Advice in Scientific or News Media

- Grants and Grant Writing (Science Careers from Science magazine, AAAS): http://sciencecareers.sciencemag.org/career_magazine/previous_issues/articles/1999_09_24/nodoi.1136734103771543891
- NIH R01 Toolkit (Science Careers from Science magazine, AAAS): http://sciencecareers.sciencemag.org/career_magazine/previous_issues/articles/2007_07_27/caredit_a0700106
- A Guide to NSF Success (Science Careers from Science magazine, AAAS): http://sciencecareers.sciencemag.org/career_magazine/previous_issues/articles/2007_07_27/caredit.a0700108
- NSF Grant Reviewer Tells All (Science Careers from Science magazine, AAAS): http://sciencecareers.sciencemag.org/career_development/previous_issues/articles/2310/nsf_grant_reviewer_tells_all/
- Research funding: 10 tips for writing a successful application (Guardian newspaper): http://www.theguardian.com/higher-education-network/blog/2013/apr/19/tips-successful-research-grant-funding
- Beginnings — How to write your first grant proposal (Nature — SoapBox Science blog): http://blogs.nature.com/soapboxscience/2012/07/31/beginnings-how-to-write-your-first-grant-proposal
- Funding: Got to Get a Grant (Nature Jobs, Nature): http://www.nature.com/naturejobs/science/articles/10.1038/nj7385-429a

Example Proposals

- NIH — NIAID: Samples and Examples
  - http://www.niaid.nih.gov/researchfunding/grant/pages/samples.aspx

Advice on Specific Topics

Advice on Impact

- Research Councils UK
  - ○ http://www.rcuk.ac.uk/ke/impacts/
- SFI:
  - ○ http://www.sfi.ie/funding/sfi-research-impact/impact-statements/ what-makes-a-good-impact-statement.html
- NSF Broader Impacts:
  - ○ http://www.nsf.gov/od/iia/publications/Broader_Impacts.pdf
  - ○ http://www.nsf.gov/od/iia/special/broaderimpacts/?WT.mc_id= USNSF_51

Presentations by Agency Officials

- ○ https://hvcfp.net/wp-content/uploads/2014/07/Grant-Writing-Skills.pdf
- ○ http://www.sfi.ie/assets/files/downloads/Funding/Funding%20 Calls/pi/grant%20writing%20slides.pdf

Resources/Presentations by Universities/University Researchers

https://doresearch.stanford.edu/research-scholarship/about-proposals/ successful-proposal-writing

- ○ http://www.gla.ac.uk/media/media_146420_en.pdf
- ○ http://www2.warwick.ac.uk/services/rss/funding/apply_funding/prep_ research_proposal/resources/succ_grant_app2011.ppt
- ○ http://www.cardiff.ac.uk/racdv/preaward/Writing%20a%20Grant/bro-chure%20dec%202013.pdf
- ○ http://www.cs.cmu.edu/~sfinger/advice/advice.html
- ○ http://www.ifeh.org/afa/docs/Writing%20research%20grant%20pro-posals%20-strathclyde%2015032011.pdf
- ○ https://courses.physics.illinois.edu/phys598pen/lectures/Phys598 ProposalWriting_SP14.pdf
- ○ http://ls.berkeley.edu/graduate/grant-writing-resources

References

Chapter 2

1. See for example at http://www.sciencebuddies.org/blog/2010/02/a-strong-hypothesis.php.
2. Franklin Berger, Biology Department, University of South Carolina, private communication.

Chapter 8

1. **DJEI (2012)**. *'Intellectual Property Protocol: Putting Public Research to Work for Ireland: Policies and Procedures to Help Industry Make Good Use of Ireland's Public Research Institutions'*. A Publication of the Irish Government's Department of Jobs, Enterprise and Innovation.
   http://www.enterprise.gov.ie/Publications/Intellectual_Property_Protocol_Putting_Public_Research_to_Work_for_Ireland_PDF_2_12MB_.pdf. Accessed on 15 February, 2015.
2. **Montobbio, F. (2009)**. *'Intellectual Property Rights and Knowledge Transfer from Public Research to Industry in the US and Europe: Which Lessons for Innovation Systems in Developing Countries'*, in *'The Economics of Intellectual Property: Suggestions for Further Research in Developing Countries and Countries with Economies in Transition,* World Intellectual Property Organization, pp. 180–209.
   http://www.wipo.int/export/sites/www/ip-development/en/economics/pdf/wo_1012_e_ch_6.pdf. Accessed on 15 February, 2015.

Chapter 9

1. **OMB (2012)**. '*A Guide for Indirect Cost Rate Determination: Based on the Cost Principles and Procedures Required by OMB Circular A-122 (2 CFR Part 230) for Non-profit Organizations*'.
Available at: http://www.dol.gov/oasam/programs/boc/costdetermination-guide/cdg.pdf. Accessed on February 15, 2015.
2. **Gilbert, L. (2007)**. '*Preparing a 'Full Economic Costing' Budget for a JISC Research and Development Proposal*'.
Available at: http://www.jisc.ac.uk/media/documents/funding/project_man-agement/fecguidance.pdf. Accessed on February 15, 2015.

Chapter 11

1. **Ermini, L., *et al*. (2008)**. *Complete Mitochondrial Genome Sequence of the Tyrolean Iceman*. Current Biology, 18(21), pp. 1687–1693.
2. From http://www.sfi.ie/funding/funding-calls/closed-calls/sfi-career-development-award-programme-2013.html. Accessed on May 7, 2014.
3. Information on both SBIR and STTR programs available at: http://archive.sba.gov/aboutsba/sbaprograms/sbir/sbirstir/index.html.
4. US Government SBIR/STTR Program. Available at: http://www.sbir.gov/about/about-sbir.
5. A summary of advice on preparing and SBIR/STTR proposal is given by Gregory Milman of NIH and can be found at: http://www.niaid.nih.gov/researchfunding/sb/documents/writerevs.pdf.
6. A Youtube video by Ruth Shuman at NSF provides information and advice on SBIR/STTR programs. Available at: https://www.youtube.com/watch?v=dPLd4_5IVFY.

Appendix 1

1. There are many commercial companies that provide electronic grant management. A list of 13 such companies provided by Foundation Centre is available at: http://foundationcenter.org/grantmakers/e-grants_instructions.html.

Endword

So here we are at the end of the book. You should now be in fine shape to tackle writing an excellent research proposal. We hope that you have found this book helpful, informative and enjoyable.

We trust that you will have learned many useful skills about making your proposal exciting, clear and responsive to the goals of your funding agency. You will be ambitious but not too ambitious. You will know to begin preparing your proposal well before the deadline and to leave time for a critical review of your proposal both by yourself and others before submission. You will be aware of the importance of spelling out your methodology clearly and with enough detail to be convincing to the reviewers. You will be seeking to show the positive impact of the research you are proposing and will be seeking collaborators as appropriate to strengthen the needed expertise.

Unfortunately, a very good product is still no guarantee of success. Even Nobel Laureates have proposals rejected. Successful proposal writing, like good research, requires patience and especially persistence. The adage 'if at first you don't succeed, try, try again' is particularly apt when applied to writing a grant proposal. The lesson is not to be discouraged, but to learn from a rejection and to use the feedback from the reviewers to write an even better proposal.

We challenged each other to come up with a limited number of **suggestions** to end on and here they are:

DO read the CFP carefully more than once.

DO leave plenty of time before the deadline to start writing your proposal.

DO make sure you are not the only reader of your proposal.

DO explain the **What,** the **Why**, and the **How** of your proposal right up front to capture the interest of the reviewer.

DO explain the methodology of your proposal in sufficient detail to convince the reviewers that it will succeed.

DO check the spelling and grammar of your proposal carefully so as not to give the reviewers a bad impression.

Index